DOING WITHOUT
THE
PHOTOCOPIER

From A to Z: 26 Creative Ideas for Reusable Language Games and Activities

By Elaine Kirn
West Los Angeles College / Authors & Editors

DOING WITHOUT THE PHOTOCOPIER
From A to Z: 26 Creative Ideas for Reusable Language Games and Activities

FIRST EDITION

9 8 7 6 5

Copyright © 1995 by Authors & Editors

AUTHORS & EDITORS
Elaine Kirn-Rubin / Arthur Rubin

10736 Jefferson Boulevard, #604

Culver City, California 90230

Tel. (310) 836-2014 / FAX (310) 836-1845

e-mail: info@2learn-english.com

www.authorsandeditors.com

Published in the United States of America by Authors & Editors

ISBN 0-9627878-4-1

Manufactured in the United States of America

TABLE OF CONTENTS

READING SKILLS

WRITING SKILLS

VOCABULARY BUILDING

CONTENT (SUBJECT MATTER)

CULTURE

DOING WITHOUT THE PHOTOCOPIER
From A to Z: 26 Creative Ideas for Reusable Language Games and Activities

A Letter to Colleagues

DEAR LANGUAGE-TEACHING SPECIALISTS:

We all know that photocopying pages for student use is time-consuming, expensive, and often illegal! Yet creative instructors usually want to supplement textbook lessons--perhaps to adapt a one-level text to a multi-level class, to provide reinforcement and additional practice (in a different form) of textbook material, to offer the variety and challenge of a change of pace, or to provide more opportunity for student-centered activity. There are always reasons to go beyond the text.

WOULD YOU LIKE TO SAVE TIME, MONEY, AND ENERGY?

Because I wanted to supplement the textbooks my students had purchased without violating my own or other authors' copyrights, I began creating materials in reusable "kit" or "set" form. I soon found that I was not only reducing our department's photocopying bills but that I was also saving myself and my colleagues a considerable amount of time in lesson preparation: once an activity had been created and class tested, it could be reused (1) in the same class on

another day, (2) in another section of the same course, (3) in the same course in another semester, and/or (4) to supplement a related course. For example:

⇒ *"Information Bingo" (Idea H) is an all-purpose getting-acquainted activity that always works as a "warm up" to any communicative content area. In various forms, it can be reused to introduce a new unit of study, such as "work," "citizenship," "interests," "fields of study," and many others.*

⇒ *"The Homophone Game" (one version of Idea S--"Word Matching") has been equally successful at both low and high intermediate levels of community college vocabulary courses. It has also been used effectively as a spelling activity in writing courses.*

⇒ *"The Expert Game" (Ideas V and W) has always been well-received--at both the intermediate and advanced levels of reading courses, as well as in writing and speech courses--because it involves all language skills, continual student interaction, and immediate feedback on language performance.*

DO YOU NEED IDEAS FOR A VARIETY OF TEACHING SITUATIONS?

The ideas in this "Doing without the Photocopier" resource book are uniquely creative and flexible in their application to the language classroom--ESL (English as a Second Language), ENL (English as a Native Language), Basic Skills (sometimes called Developmental Communications), foreign languages, and related areas. When used with native speakers, various ideas may be suitable to the K through 12 environment as well as the adult-school and community-college classroom. If second-language learners are involved, only some of the ideas will be useful in elementary schools, but the others are suitable for secondary-school, adult-school, college, and university ESL courses and programs as long as the students are not rank beginners.

Whereas the suggested activities in many idea books for instructors and trainers may be designed for one purpose only--to teach a specific grammar point, a particular group of vocabulary items, a learning principle, or the like--the ideas in "Doing without the Photocopier" are much broader. Each idea is meant to serve as a teacher-training or self-training concept. For clarity, the examples for each set of instructions are based on a specific context, but the suggestions for adaptation to different levels of language proficiency, the possible variations, and the list of other areas of application are designed to stimulate your creative juices. As you try your own version of each activity, you will learn a lot about teaching and learning. You (and/or your students) will develop unique classroom techniques that work optimally in your specific situation(s).

DO YOU FOCUS ON SPECIFIC LANGUAGE SKILLS IN YOUR TEACHING?

The twenty-six ideas in "Doing without the Photocopier" are divided into eight areas of language focus:
⇒ SPELLING AND PRONUNCIATION
⇒ GRAMMAR
⇒ LISTENING AND SPEAKING
⇒ READING SKILLS
⇒ WRITING SKILLS
⇒ VOCABULARY BUILDING
⇒ CONTENT
⇒ CULTURE

If any of your courses are limited to one or two language skills (such as listening and speaking only), first try the ideas that focus on those areas (e.g., Ideas H-K for an Oral Skills course), omitting the steps that demand other skills. Other ideas will be useful in the same courses if you adapt the steps; for example, you could substitute oral presentations for composition writing.

No matter what kinds of students you have in your language class, you can engage and motivate them with these creative activities.

DO YOU USE AN INTEGRATED LANGUAGE SKILLS APPROACH?

Most ideas, nevertheless, involve skills or content beyond the language focus area. For instance, grammar activities include listening/speaking or writing steps. The "Expert Game," listed under the category of "Content and Culture," provides practice in several oral and written language skills, including the ability to take notes. The purpose of the above categorization is to provide a reference point for your choice of games or activities. Keep in mind, though, that nearly all ideas (especially those listed under "Vocabulary Building," "Content," and "Culture") offer opportunities for students to practice any or all of the four language skills--reading, writing, listening, and speaking.

CAN YOU FACILITATE STUDENT-CENTERED LEARNING EFFECTIVELY?

I believe in individualized as well as student-centered and cooperative learning. Thus each idea in this book calls for active participation by students at every stage of the game or activity. After you have prepared, set up, and demonstrated each step, you will be free to facilitate the activity, giving individual attention to students at whatever level they are capable of working. You can observe and learn from student interaction. The "pressure is off"--you will rarely be called on to "perform" before a large group of students with different levels of ability, varied needs, and diverse interests. Best of all, because all students will be involved actively and simultaneously in the learning process, the "tedium factor" will disappear. Gradually, students will learn to help one another and to take responsibility for their own achievement.

ARE YOU LOOKING FOR WAYS TO ENJOY TEACHING MORE?

I hope the ideas in "Doing without the Photocopier" enliven your methodology and motivate your students as they have mine. I hope these offerings heighten the effectiveness of your teaching. Most of all, I hope you enjoy using these ideas. Feel free to share them with your colleagues and to improve on them in ways that work best for you and your particular classroom style.

I invite your reactions, comments, and suggestions.

E.K.
Culver City, CA

"This teacher-resource book and its sequel, Still Doing Without the Photocopier, were designed for experienced or adventurous, creative instructors and tutors who want to go beyond the ordinary in their teaching effectiveness. Thank you for being out there—and enjoy!"

E.K.

ABBREVIATIONS USED IN THIS BOOK:

S = student   Ss = students
s/he = he or she

Students may blame their pronunciation problems on the mistaken belief that English words aren't pronounced the way they're spelled. To practice the relationships between sounds and spellings (phonics), try this versatile game. It is effective because players get immediate feedback on the comprehensibility of their pronunciation.

⇒ **SPECIFIC CONTENT OF THESE INSTRUCTIONS: the most common spellings of ten vowel sounds: / æ ɛ ɪ ɑ ə e^y i^y a^y o^w u^w /.**

⇒ **MATERIALS: For each group of six students: (a) five Bingo boards, numbered 1-5, each consisting of the same 16 words in different order (b) one set of 16 small word cards for the "caller" (c) small markers—such as squares of paper or dried beans.**

Some examples of *Phonics Bingo Cards* available from Authors & Editors. These instructions explain how to make simpler cards of your own—or to get learners to produce their own cards as a learning activity. On the right are ideas for possible kinds of Bingo markers.

INSTRUCTIONS

1. Divide the class into groups of up to six Ss each. Distribute one set of five different "Bingo boards" to each group. Each S works on a different board, with two Ss sharing one board for the introductory game.

2. Distribute Bingo markers. So that Ss can hear the correct pronunciation of the words, play a model game with the whole class: using one of the decks of word cards, call out the words one by one. Ss cover the words with markers. (If they are playing correctly, one S from each group will have "Bingo" at the same time.)

Teaching Tips

Before they begin the game, remind players to keep the materials intact. After the game, each group is to return a set of five different Bingo boards, put the markers in a designated container, and put a rubber band around the set of word cards.

3. Give each group a deck of sixteen word cards.

4. Each group chooses one S as the first "caller." This player shuffles the caller cards. S/he calls out the words from the cards in order, attempting to make the spelling clear from his/her pronunciation alone. (If the group doesn't seem to understand, that person can repeat a word, use it in a sentence, or even spell it aloud, if necessary, but s/he should not *show* a card to the group.) The caller places the "used cards" in a separate pile.

5. Groups play the game at the same time. Play in each group continues until one S has covered a "Bingo" line--i.e. four adjacent horizontal, vertical, or diagonal squares-- and calls out "Bingo!" That player wins the game if s/he can pronounce the words and if they are truly words that the caller has called.

6. You might want to provide small prizes. Players empty their Bingo boards and start a new game. The winner of the last game becomes the new caller.

7. For follow-up, play the same game with the entire class. Each S in turn calls out a different word of his/her choice; continue until all the boards are completely covered. Review word meanings and uses, if desired.

Teaching Tips

If a player wins more than one game, then another player should become the next caller. Ideally, the game should be played five times so that each S is the caller once. Or—to make the game move faster, try one of the alternative instructions on the next page.

8. Collect all materials. Arrange them for the next game session.

beat
bite
boat
bit
bat
bait
bought
but
bet
boat
boot

LEVELS = BEGINNING TO INTERMEDIATE
⬃ SUGGESTIONS FOR ADAPTATION ⬂

⇓ In classes of beginners or of students from different language backgrounds who have trouble understanding one another, callers can provide "hints" to insure student success: (a) After calling out a word two or three times, they tell the meaning. Or the other players--to make sure they are covering the right square--can tell what they think the meaning of the word is. (b) After calling out the word several times, callers can spell the word aloud or show the card. (c) Players can check that they are covering the same word as other players.

⇔ Instead of one player being the caller for each game, the caller cards are shuffled and distributed equally at the beginning of each game. Each player in turn calls out a word. Or dispense with caller cards altogether. Instead, Ss in turn choose words to call out from those that are uncovered on their boards.

⇑ If you have produced several *different* sets of Bingo games—to practice different groups of sounds, give one set to each group of Ss. After they have finished playing that game—if they have mastered those sound contrasts—they can trade sets with another group.

⇑ Higher-level classes can make their own "Bingo boards." First, decide if the boards should consist of 16, 25, or even 36 squares each. Then, from one deck of caller cards, pronounce the words one by one; individually, each S writes each word in a different rectangle of his/her board *in random order*. (Walk around the room to check that the boards are coming out differently.) A few seconds after calling out each word, have Ss spell it aloud and/or write it on the chalkboard to insure that the boards are correct.

OTHER AREAS OF APPLICATION: Other groups of contrasting sounds, such as other vowel sounds, initial consonants, and ending consonants. At higher levels, include words with less common or more unusual spellings for the sounds being contrasted (e. g. pace / peace / piece; she / shy / show / shoe) as well as confusing word groups (e.g., advise / advice; please / police; we're / were / where / wear / weird).

RELATED AUTHORS & EDITORS MATERIALS: Prepared *Phonics Bingo Games* are available on four levels of difficulty. Each level includes four different games for vowel sound contrasts and four different games for consonant sound contrasts (initial and final). There are ten Bingo boards for each game—with pictures for Levels A-C—and one set of caller cards. For more information, see the Authors & Editors order form at the end of this book.

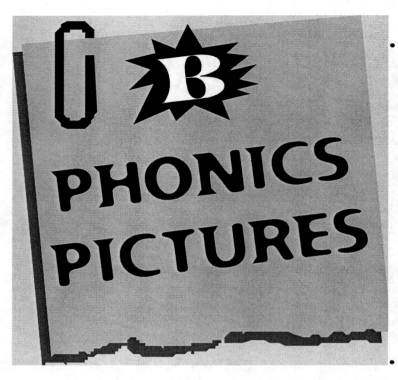

PHONICS PICTURES

Students enjoy vocabulary lessons, especially activities based on pictures. Why not combine vocabulary with instruction and practice in pronunciation, phonics (the relationships between sounds and letters), and spelling?

⇒ **SPECIFIC CONTENT OF THESE INSTRUCTIONS: Initial Consonant Sounds / b d ch f g h j k l m n p s sh t v w y z /.**

⇒ **MATERIALS: Large complex pictures, probably scenes, containing many different items and/or activities--several more pictures than there are groups; large paper to write on.**

You can find big picture scenes—including funny ones—in magazines, children's books, beginning language texts, and so on. This cartoon illustration is from an old calendar.

INSTRUCTIONS

If necessary, review the sound-spelling correspondences you have taught—in this case the spelling of initial consonant sounds.

1. On the chalkboard, draw an empty grid for Ss to copy. At the top of the grid, write the sounds to be practiced, one symbol for each column.

2. Ss divide into teams (groups) of about four people each. On a large piece of paper, each team copies the format of the grid from the board. Give one picture to each group, and put the extras on a table in front of the room. Set a time limit for groupwork.

/b/	/ch/	/k/	/s/
<u>b</u>eans	<u>ch</u>air	<u>c</u>ap	<u>s</u>igns
<u>b</u>oy	<u>ch</u>ecke	<u>c</u>andy	<u>s</u>ausage
<u>b</u>oxes	<u>ch</u>eek	<u>c</u>lock	<u>s</u>tool
<u>b</u>askets	<u>ch</u>ild		<u>s</u>mile
<u>b</u>arrel			

3. Team members work together to write the words from their picture in the appropriate columns of their grid. (Remind Ss that they are to list the words in columns according to *sounds*, not letters.) As soon as group members have written down as many words as they can think of, they exchange their picture for another on the table and continue working until time is called.

4. Ss stay in their teams to compete with other groups. For each sound category, one member of each group tells one word in turn. Write the word on the board (in the appropriate list on the chart) as that group member spells it aloud. If other teams have the same word, they cross it off their list so that it will not be repeated.

5. The winner for each category of sounds is the team with the largest number of correctly-spelled words (or the largest number of words that no other group has thought of). Provide a small prize that can be shared by group members. The "grand-prize winner" is the group with the largest total number of correctly-spelled words (or the largest total number of words that no other group has thought of) in all columns--all the words collected from the pictures.

6. For follow-up, review the meanings of the words on the board. Point out how they conform to or are exceptions to the learned spelling patterns. On another day, use them for a spelling quiz.

LEVELS = HIGH BEGINNING TO HIGH INTERMEDIATE
↙ SUGGESTIONS FOR ADAPTATION ↘

⇓ For beginners, list relevant vocabulary from the pictures on the board. Go over the meanings. Erase or cover the words before beginning the game.

⇓ To make the activity easier, list a different *letter* at the top of each column (instead of a symbol for a sound). Ss list words beginning with that letter regardless of which sound it produces. For example, <u>c</u>ake and <u>c</u>ircle would both appear in the "c" column. <u>K</u>ey and <u>k</u>not would be in the "k" column).

⇓ In low-level classes, each group can use the *same* picture, perhaps a scene from the textbook. Ss can work in pairs or individually instead of in groups.

POSSIBLE VARIATIONS

⇔ Instead of using pictures, play the game of "Categories." Have each group copy a grid of rectangles

Categories	/b/	/ch/	/k/	/d/	/j/
animals	<u>b</u>ear	<u>ch</u>eetah	<u>c</u>at		<u>g</u>erbil
objects	<u>b</u>aseball	<u>ch</u>ain	<u>k</u>ey	<u>d</u>ime	<u>j</u>ar
people	<u>b</u>oy			<u>d</u>octor	
actions		<u>ch</u>op			<u>j</u>ump
adjectives	<u>b</u>oring		<u>c</u>old		

rather than columns. In addition to writing a phonetic symbol or letter above each column on the grid, they write a category to the left of each row—such as animals, objects, places, foods, job titles, people, etc. Higher-level groups may suggest more specific categories, such as kinds of furniture, clothing, things in a classroom, colors, and the like. And if your Ss are in similar lines of work, the grid can contain vocational words, such as names of tools, computer terms, work activity verbs, etc.

Groups work together to be the first to copy and fill in the grid. The chosen words must not only contain the sound at the top of each column but also conform to the category to the left of each row. Score as described above.

OTHER AREAS OF APPLICATION: Vowel sounds or letters, final consonant sounds or letters.

To master spelling, students must be able to write the words they hear and pronounce the words they can spell. This activity provides practice and immediate feedback in both skills: spelling through phonics and reading words aloud—comprehensibly.

DYAD SPELLING

⇒ **SPECIFIC CONTENT OF THESE INSTRUCTIONS:** Vowel sounds spelled by one vowel letter or by two or more letters together.

⇒ **MATERIALS: Writing paper.**

LIST 1	LIST 2
1. h__ea__r	1. h_____r
2. l__oo__se	2. l_____se
3. str__ee__t	3. str_____t
4. ch__ai__r	4. ch_____r
5. b__uy__.	5. b_____.
6. f__u__n	6. f_____n
7. p__ou__r	7. p_____r
8. tr__ee__.	8. tr_____.
9. l__igh__t	9. l_____t
10. r__u__b	10. r_____b

An example of a *Dyad Spelling* paper.

I *NSTRUCTIONS*

If necessary, review sound-spelling relationships—in this case the spelling of medial consonant sounds (excluding spellings with final silent *-e*).

1. On the board, list numbered incomplete words—with blanks replacing the letters that will illustrate the spelling principle being covered. There are examples on the previous page.

2. Ss fold their papers in half lengthwise and copy the word list from the board twice, once in each column. In each blank of the left column, they fill in the letter or letters for the missing vowel sound. Make clear that they are to create real words of one syllable only and that there are several correct possibilities. (You might remind them that a few silent *consonant* letters can be part of the spelling for vowel sounds, as in *ni<u>ght</u>* or *strai<u>ght</u>*.) For example, item 1 in List 1 on the previous page could be *hear, hair, her, or hour*. Item 2 could be *loose, lose, lease, or louse*.

Teaching Tips

If you are teaching phonics principles and working on vowel sounds, remind Ss to write letters for vowel sounds only—even if *consonant letters* are included, as in the combination *ou<u>gh</u>*. Otherwise, they are likely to add consonant sounds or even extra syllables.

3. When Ss have completed their lists, they pair up. Each S in turn reads aloud his/her words to his/her partner; without looking at the first S's paper. The second S writes the missing letters in the words in the right column. Then they reverse roles and repeat the process.

4. Partners compare papers. If they have pronounced and spelled their words correctly, and if their partners were able to understand and spell those sounds, the left column of the first S's paper should be identical to the right column of the second S's paper, and vice versa. If there are discrepancies, Ss circle them and figure out how the mistake occurred--through a spelling error, inaccurate pronunciation, etc.

5. On the board, Ss list the words that presented problems. Go over these (spelling, phonics rules, pronunciation, and meaning) with the class. Have a S copy the list of words on paper for you. Other Ss may wish to make copies for themselves.

6. For follow-up on another day, use the spelling list (or other words illustrating the same principle) for a spelling quiz based on the above techniques. To begin, list words with blanks on the board. Each S copies this list on his/her paper. Second, say the words you have in mind aloud. Ss fill in the missing letters on the basis of your pronunciation alone. Third, repeat the words in context--each word in a phrase or sentence. Ss correct their work if necessary. Finally, ask the class to read you the words aloud. Fill in the missing letters on the board. Ss correct their own papers. Collect scores if desired.

C. Dyad Spelling

LEVELS = HIGH BEGINNING TO HIGH INTERMEDIATE
↙ *SUGGESTIONS FOR ADAPTATION* ↘

⇓   For the lowest levels, prepare two pages based on the same sounds. On one of the papers, fill in letters in the left column in one way and on the second page, in another way. Each S in each pair receives a different paper. The first S reads his/her words aloud to his/her partner, who fills in letters for the sounds s/he hears in the right column of his/her paper, and vice versa. S pairs compare papers.

Teaching Tips

To save paper, you can put a different "Dyad Spelling" exercise on the back of each page. Alternately, half the class could copy a word list from one chalkboard and the other half, from another. An idea that would require more advanced preparation is to laminate the pages so that Ss could write on them and then erase their marking.

POSSIBLE VARIATIONS

⇔   Instead of simply filling in letters, Ss in pairs can make up their own spelling lists and dictate the words to each other. If you try this variation, you might want to supply a list of word pairs or groups to choose from (e.g., *pat, pet, pot, put, putt; pair, peer, pour, poor,* etc.).

⇔   Instead of spelling lists, you can supply simple paragraphs with letters (or whole words) to fill in. Each S "dictates" an entire paragraph to his/her partner, who writes the words and then checks his/her work.

⇔   Just give each person in a pair a different paragraph to dictate to his or her partner. Then everyone compares what he or she has written to the original.

One person dictates the paragraph to another. The second writes the missing letters in the blanks. Then they compare papers. Are they exactly the same? If so, the pronunciation was probably good. And the other learner can listen and spell well.

If the two papers are different, what's the problem? Is it accent or vocabulary or spelling? Figure it out together.

One p__rson d__ctates the paragr__ph to an__ther. The s__cond wr__t__s the m__ssing letters in the bl__nks. Th__n they comp__r__ p__pers. Are they ex__ctly the s__m__? If so, the pron__nciation was pr__bably g__d. And the __ther l__rner can l__sten and sp__ll w__ll.

__f the two papers are d__fferent, what's the pr__blem? Is it __ccent or voc__bulary or sp__lling? F__gure it __t tog__ther.

OTHER AREAS OF APPLICATION: "Complex" or "long" vowel sounds spelled with final silent e; initial or final consonant sounds, including consonant clusters.

SPELLING GRIDS

Good spellers develop a feel for the language they are writing, not only by being able to sound out words but also by becoming accustomed to common letter combinations. Help your students to acquire this useful skill by playing this simple pencil-and-paper spelling game.

⇒ **SPECIFIC CONTENT OF THESE INSTRUCTIONS: None.**

⇒ **MATERIALS: Graph paper with relatively large squares or grids prepared beforehand--cut into rectangles of about 4" x 5" or larger.**

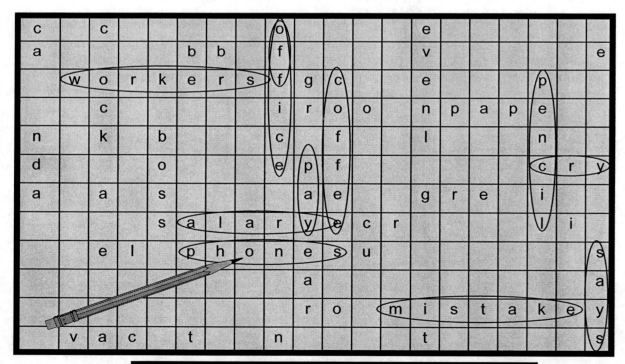

An example of a one player's Spelling Grid, partially filled in.

INSTRUCTIONS

1. Hand out (or have Ss prepare) grids. If necessary and desired, demonstrate the activity by playing one game with the class, on the chalkboard or on the overhead projector. Then divide the class into small groups to play the game. All groups play at the same time.

2. The first player in each group calls out a letter of the alphabet. All players write that letter in any single square of their grids. The next player calls out a second letter (the same or different from the first), and all players write that letter in another square of their choice. The game continues in this fashion until all players have filled their grids with letters.

3. The object of the game is to create words from the letters, printed across or down. Once a letter has been written into a square, its position may not be changed. Each letter may appear, however, in more than one word.

Teaching Tips

Because this game includes strategy, knowing spelling patterns and rules will help players to accumulate higher scores. For example, the better spellers will know that the letter *u* always follows the letter *q*, that almost no words end in the letter *v*, that *ph* and *gh* and *wh* are common letter combinations, and so on. Therefore, they will be better able to play their placement of letters. If they also know about the spelling of prefixes and suffixes when they are added to words, they will be able to produce longer words on their grids. You might want to reveal some of these "secrets" to Ss before or as they play this game.

4. At the end of the game, the player in each group with the most correctly spelled words is the winner. Alternatively, players can receive one point for each *letter* that is included in a word. The player with the highest score wins. If desired, provide small prizes.

5. For follow-up, each S can hand in a list of his/ her words for comments on spelling, correction, and/or a grade. Ss can list their three to five "best" (or longest) words on the chalkboard to include in a brief class vocabulary and spelling lesson.

Students might want to play "Spelling Grids" endlessly.

⇓ Instead of filling out their grids while following game rules, beginning students can form words by filling out a grid on their own. The object remains the same—to create as many correctly-spelled words as possible. The only rule is that they cannot change a letter once it is written--although even this rule can be ignored. Another possibility is to have S teams compete to form words, all members of each team working on the same grid.

⇓ For lower levels, provide grids with fewer (larger) squares. To help Ss think of words, direct them to previously-taught spelling lists, reading lessons, etc.

⇑ For higher levels, grids can contain more squares. For variety, students can create grids of various shapes (oblong, L-shaped, in the shape of a cross, etc.) Players can include words printed diagonally as well as horizontally and vertically.

⇑ To provide challenge to advanced Ss, make the rules more difficult. For example, you can limit the categories of the words that may be formed (e.g., only nouns, only verbs, only words associated with work, etc.). Or you can allow them to score only words of four letters or more.

Some "novelty" Spelling Grids.

POSSIBLE VARIATIONS

⇔ Ss do not circle the words they have created, even after completing the game. Instead, they list the words on the back of the paper or on another paper to compute their score. Then they trade grids with a partner, who approaches the grid as a "word search puzzle," circling the words s/he finds. S/he compares the words s/he has found with the creator's list of answers.

⇔ Ss (or S teams) fill out a grid with the object of creating a crossword puzzle. They copy the pattern they have created, leaving out the letters, number the first square of each word, and write the corresponding clues in columns titled "Across" and "Down." Ss (or teams) exchange puzzles to complete.

Grammar lessons should be contextualized, which means that they should approximate "real life" as closely as possible. So that students can use new grammar realistically, bring pictures of everyday things and interesting situations into the classroom.

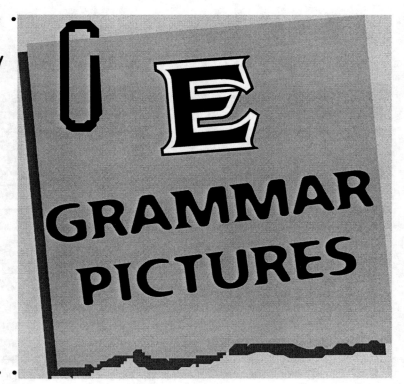

GRAMMAR PICTURES

⇒ **SPECIFIC CONTENT OF THESE INSTRUCTIONS: Modal verbs of possibility and probability (past, present, and future).**

⇒ **MATERIALS: A set of pictures well-suited to the grammar being taught or reinforced, paper to list sentences on, and perhaps an answer key.**

There are many sources of free or low-cost "Grammar Pictures," such as newspapers and magazines, old picture books, and so on. These cartoons are from the *Clickart, Famous Magazine Cartoons.* The original captions have been cut off.

*I*NSTRUCTIONS

1. Teach or review the relevant grammar, in this case modal verb phrases of possibility, impossibility, and probability as shown in this grammar chart:

MEANING	PAST		PRESENT		FUTURE	
Possibility	• may (not) have		• may (not) be		• may (not)	
		+ PAST PARTICIPLE	• might (not) be	+ VERBing	• might (not)	+ VERB
Impossibility	• could(n't) have		• couldn't be		• couldn't	
Probability	• must (not) have		• must be			

2. Show some of the pictures and discuss what is happening. Then elicit the relevant grammar by asking questions such as *What do you think may have happened? What could be happening now? What might happen next?*

3. Pointing out that the pictures are numbered, pass them around the class. Ss work alone, in pairs, or in groups to write a specified number of sentences, using the relevant grammar, about each picture. (For instance, for modals of possibility and probability, they might write one past, one present, and one future sentence for each picture.) Move around the classroom to give necessary help and make suggestions. At the end of the activity, you may

Teaching Tips

Because this particular grammar is especially difficult to elicit, you might want to give a few equivalents. For example, for the first sample picture on the previous page, you might say, "Probably a flying saucer landed." Ss would give the past time equivalent "A flying saucer must have landed." The sentence "Maybe people are running out of the crushed buildings" is equivalent to "People may be (or might be) running out of the crushed buildings." "Perhaps a monster will come out of the space vehicle." means "A monster could (or may or might) come out...," and so on.

want to collect the papers for comment, correction, and/or a grade.

4. After a specified length of time, reconvene the class, making sure that each S has at least one picture at his/her desk or table. Each S in turn returns the picture after telling his/her sentence(s). The class makes corrections and tells other possible sentences.

5. Provide review, perhaps on another day, by showing the pictures and having the class recall or make sentences about

What must have happened before this scene? What might be going on now? What could happen next?

them.

LEVELS = HIGH BEGINNING TO LOW ADVANCED
↙ *SUGGESTIONS FOR ADAPTATION* ↘

⇓   For lower-level groups, show all the pictures and have the class suggest sentences before passing the pictures around the classroom.

⇓   After Ss have had time to work on their own, distribute an answer key of possible sentences. Ss can use the key to make corrections or change their own sentences, asking for advice.

⇔   In sequential-list form, write the numbers of the pictures on the chalkboard— in the case of modal verbs of possibility and probability, perhaps 1a, 1b, 1c; 2a, 2b, 2c, etc. so that Ss can write three sentences for each picture. Alternatively, arrange the numbers in chart form--with headings for past, present, and future--so that three kinds of sentences can be written for each picture. At any time while writing their lists, Ss can come to the board to fill in the space after any particular number. After reconvening the class, go over these sentences as you show the pictures again.

What must have happened before this scene? What might be going on now? What could happen next?

Teaching Tips

There are many grammar topics that can be taught or reinforced through the use of sets of pictures; just make sure that the pictures "work"--i.e. that they are particularly suited to the relevant grammar. Of course, adapt the level of the grammar presentation to the proficiency level of the class.

⇑   Ss in higher-level groups can write a paragraph about each picture instead of separate sentences.

POSSIBLE VARIATIONS

⇔   To *present* a new grammar pattern, give one picture to every other S and tell

OTHER AREAS OF APPLICATION: *There + be +* SUBJECT (places); present continuous / past continuous (activity scenes); simple past / *used to* VERB (pictures from the past); *be about to* VERB (situations in which something is about to happen); future verb forms (pictures of possible future events,); count vs. noncount nouns (food, things, substances); prepositions (pictures with item arrangements); etc.

GRAMMAR PARAPHRASES

Students tend to use the same grammar structures over and over--often the basic structures that they learned as beginners. To encourage the use of more "advanced" grammar, you can specify the structures to be used in paraphrasing.

⇒ **SPECIFIC CONTENT OF THESE INSTRUCTIONS: The past tenses, including *used to* + verb; nostalgic letters.**

⇒ **MATERIALS: At least ten different letters, paragraphs, or "news-clip" type articles of approximately the same length that lend themselves to paraphrases containing the relevant grammar—pasted up on index cards; writing paper.**

Teaching Tips

If you can't find enough *different* readings that are appropriate for this activity—or if the class is very large, you can have Ss work in pairs on the same reading.

LETTER 1

My grandparents lived in a small house next to a forest, and we used to visit them on weekends. I remember the smells of that lake and the feel of the air, especially in the humid summer. My brother and I used to catch insects--grasshoppers and lightning bugs--and study them inside jars with holes in the top.

Of course, I remember the mosquito bites, too--and the chiggers. Did they itch! But after a cold bath, my grandmother would put lotion on the bites. I still didn't feel very comfortable, but her homemade chocolate chip cookies always cheered me up.

LETTER 2

I was born in Vermont, and one of my favorite childhood activities was going to the forest when the sap was running. As soon as my father had tapped a tree, I held a bucket under it and watched it slowly fill up with sap. I loved tasting the sap and getting real maple syrup. Even today, I can't eat pancakes or waffles without the real thing.

Blueberries were another favorite. We children used to pick them on our own--and then mother made blueberry pies, blueberry pancakes, and other delicious treats.

These sample letters for practice in the past tenses are similar to letters from the magazine *Reminiscences.*

that S one or more correct sentences, using the relevant grammar, about the picture. That S shows the picture to the next S, repeating the sentences or paraphrasing them. The process continues around the circle or room.

*I*NSTRUCTIONS

1. Review the relevant grammar if necessary--in this case ways to express past activity (the simple past, past continuous, past perfect and past perfect continuous tenses, *used to* + VERB, *would* + VERB). Use one of the selections as a sample: read it to the class and help Ss paraphrase the main ideas, using the relevant sentence structures when natural. You might want to write--or have a S write--a sample paraphrase on the board.

2. Each S receives a selection. Allow time for Ss to write their paraphrases on their own. If more than one S has the same selection, they compare their summaries. Did they both (or did all three) write approximately the same ideas? If not, why not? Are all the paraphrases "correct"?

3. Ss form pairs or small groups, each with a different selection to tell about. In turn, they orally summarize their selections for their partner or group, using the relevant grammar. To retell what they understood, listeners paraphrase the main ideas.

4. If desired for additional practice, Ss write a summary of one or more of their classmates' selections, comparing what they wrote with both their classmates' writing and the original selections. You may want to collect some or all of the summaries for comment, correction, and/or a grade.

5. If you are using this activity in a writing course (and the content lends itself to

If the reading material you find includes illustrations—photos or drawings—paste them up too! They will help Ss paraphrase the general ideas without looking at the readings all the time.

student writing), have Ss write their own paragraphs, letters, or compositions like the models. Follow up with usual writing-as-a-process activities and/or with activities similar to those described above.

LEVELS = INTERMEDIATE TO HIGH ADVANCED
↙ SUGGESTIONS FOR ADAPTATION ↘

⇓ For lower-levels, simplify the authentic material before pasting it up on cards. Provide "paragraph frames" for writing and/or a model on the board that Ss can change to fit the information in their own selections.

⇓ If the class does not include composition writing, you can have Ss write lists of sentences rather than whole paragraphs--one or more sentences for

1. Simple Past: *When I was a child, I lived in Vermont*
2. Past Continuous: *When the sap was running, we often went to the forest.*
3. *Used to: We used to help our parents tap trees.*
4. *Would: The children would carry the buckets.*
5. Past Perfect: *After we had filled a bucket, ...*

each of the patterns being practiced. The sentences do not have to fit together well in sequence as they would if they were part of a well-constructed paragraph. *Examples*:

⇑ Advanced students will profit more from this activity if they write their paraphases without looking at the original selection, later checking what they have written. Alternatively, you can read or tell a story aloud and then have Ss write a paraphrase of what they understood, using the relevant grammar as often as appropriate.

POSSIBLE VARIATIONS

⇔ If the selections are informational, especially if you are using this activity in a reading or listening / speaking class, you may want to follow up with the "Whole-Class Expert Game" (Idea W). Ss in turn tell the class about their topics, and the class takes notes. Give a brief oral or written quiz on the information conveyed. Have a follow-up discussion, if appropriate.

*O*THER AREAS OF APPLICATION: To practice modal verbs and other expressions of advice, paragraphs of a "how-to" article, such as "How to Improve Your Study Habits" or "How to Advance at Work"; To practice adverb clauses (clauses beginning with *when, before, after, if*, etc.), instructions, such as "How to Play Checkers," or recipes; to practice other kinds of grammar, many other kinds of selections, all at an appropriate level of difficulty for the class.

Especially at the intermediate level and above, students differ greatly in their grammar strengths and weaknesses. Why not customize grammar "instruction" during classtime through individualized grammar activities?

⇔ If the selections are personal or fictional, use them as the basis of a discussion or other follow-up activities.

⇒ **SPECIFIC CONTENT OF THESE INSTRUCTIONS: None.**

⇒ **MATERIALS: Various ESL and native-speaker grammar textbooks no**

Teaching Tips

Only those textbooks that divide the language into discrete grammar areas and points (such as verb tenses, the passive, pronouns, etc.) and that provide exercises as well as explanations are useful for this activity. A textbook in which grammar is taught "inductively" or through a "notional-functional" approach, for example, would not be suitable.

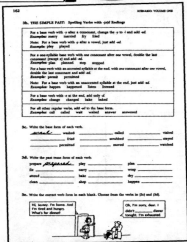

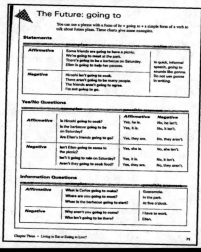

Some sample pages from old ESL grammar texts that include answer keys for self study.

longer in use by the school--those kept in a resource collection or about to be discarded, including duplicates; separate or photocopied answer keys; writing paper.

INSTRUCTIONS

Teaching Tips

If the course does not include a composition or other writing component, each S can choose (or you can suggest) the specific points that s/he needs to focus on or review.

1. After reading a typical piece of each S's writing, list a few grammar points with which he or she seems to have particular difficulty—such as articles, the present perfect tenses, infinitives, prepositions, etc.

2. When you hand back the compositions, provide a collection of grammar textbooks. Distribute these to Ss.

3. Ss work individually, in the style of a "learning lab." They find suitable

Teaching Tips

If there is not at least one book for each S, you can divide the books into parts beforehand, stapling together the pages of each lesson, chapter, or unit. In this case, Ss will have to choose sections that apply to the grammar areas they are to work on.

grammar sections to cover, study the explanations, complete the exercises on writing paper, and check their work with the provided answer key and/or consult with you or a tutor about their answers. You may want to collect their papers for comments and/or a grade.

4. Ss rewrite their original papers according to the markings and suggestions you have provided, concentrating on the grammar points they have just

GRAMMAR AREA	POSSIBLE WRITING TOPICS
• kinds of nouns: count vs. non-count; indefinite vs. definite	• your favorite foods • how to prepare a dish
• impersonal *there is / are*	• description of a scene
• the simple present vs. the present continuous verb	• Describe your daily life. *What do you regularly do? What new or interesting*
• the present perfect tenses	• *What interesting experiences have you had since you came to this country?*
• the passive voice	• how something is made
• gerunds and infinitives	• your freetime activities and interests
• pronouns	• self-esteem

studied. Alternatively, if no compositions have been written, Ss can demonstrate mastery of those points by writing a paragraph on a topic of your suggestion that would elicit them. Following are some examples of possible writing topics for grammar practice in various areas.

LEVELS = INTERMEDIATE TO HIGH ADVANCED
⬐ SUGGESTIONS FOR ADAPTATION ⬎

⇔ For lower-level groups, it might be a good idea to prepare a "grammar kit" for this activity, similar to the "reading kit" described in *Idea L* of this book. Paste up individual grammar exercises on separate cards, perhaps with the grammar explanation on the other side. You can provide answers on another card or in an answer key. Ss work on one exercise at a time, checking their answers before they go on to another.

⇔ Adapt the kinds of textbooks or exercises provided to the levels and particular skills of the Ss. For instance, lower-level Ss will do better on controlled exercises requiring precise answers, such as fill-ins, matching, words to arrange, and the like. More advanced Ss will be able to do exercises that require completion of sentences, full-sentence answers, vocabulary or information of their own, and so on.

⇑ To force higher-level Ss to make use of more advanced grammar in their compositions or paragraphs, specify exactly which patterns or points they

Because we like <u>Italian food</u>,√ we have <u>the pizza</u> × or <u>the pasta</u>× at least <u>three times</u>√ <u>a week</u>.√ For <u>family</u>× of four, we order <u>two large pizzas.</u>√ <u>My wife</u>√ prefers <u>vegetarian pizza</u>√ with <u>onions,</u>√ <u>bell peppers,</u>√ <u>mushrooms,</u>√ and other <u>vegetable</u>× . I would rather have <u>a sausage</u>× or <u>a pepperoni</u>× on <u>my pizza.</u>√ <u>The children,</u>√ however, like <u>pizza</u>√ with <u>ham</u>√ and <u>cheese</u>√ ....

√ = correct; × = error. *Score: 13 correct - 6 errors = SCORE of 7*

Even a pizza can be the subject of a grammar-writing activity.

INFORMATION BINGO

Information Bingo is a classic Listening/ Speaking game that always works! Use this effective communication tool several times during the course--not only to help students get acquainted but to "set the tone" of cooperation and cooperative learning.

are to include. Then use this method of scoring: give one point for each pattern used correctly and subtract one point for each error in the relevant grammar. Here is an example in a paragraph about food, using count and non-count nouns—with the relevant grammar underlined.

Nacharee Coletti Thailand manicurist	Mickey Maus Austria actor	Carlos Alessio Italy graphic artist	Peter Pan U.S.A. dancer
Kazuhiro Ito Japan manager	Martin Bratiluva Czech Republic athlete	Laura Esquivel Mexico writer	Ricardo Ricardo Cuba musician
Wolfgang Pock Germany cook	Chien Chin Chu China doctor	Michael Faye Singapore student	Sheerluck Homes England detective
Elaine Kirn Switzerland chocolate	Arturo Rubén Uruguay race car driver	Uri Gargarin Russia pilot	Edson Pelé Brazil businessman

A sample "Information Bingo" Board—useful for getting acquainted.

⇒ **SPECIFIC CONTENT OF THESE INSTRUCTIONS: "Getting-acquainted small talk."**

⇒ **MATERIALS: 8 1/2 x 11" (or larger) paper; small prizes.**

INSTRUCTIONS

1. Teach or review questions and answers for getting personal information: decide or have class decide on 1 to 5 questions useful in meeting people for the first time. List these on the chalkboard. Here are possible examples:

Teaching Tips

> If the class is smaller than 16 students, they fold paper into 9 equal squares--three rows of three squares each; in contrast, in a large class of fairly advanced students, each "Bingo board" can consist of 25 or even 36 squares.

   1. What's your name? (How do you spell it?)
   2. Where are you from?
   3. What do you do (for a living)?

2. Ss fold paper into 16 equal squares--four rows of four squares each.

3. Ss circulate around classroom, gathering information from 16 different classmates (and perhaps the instructor) by asking the questions they have agreed on. They write information about each person in a different square of their *Bingo Boards*.

4. The class reconvenes. Review the rules of *Bingo*. Introduce one person to the class by telling the answers to the getting-acquainted questions.

   Example: "Watcharee Coletti is from Thailand. She's a manicurist."

5. All Ss with information about that player put an X on the corresponding square of their *Bingo Board*. That S introduces a classmate, all players with that information put an X on the corresponding square, the introduced S introduces another person, and so on until one or more players have "Bingo" (four X's in a row--across, down, or diagonally). The player with "Bingo" introduces all people in his/her Bingo row, column, or line. If s/he is correct, s/he wins a small prize. So that more players can be introduced, play continues until more Ss have "Bingo." They reintroduce those in their Bingo lines and win a prize.

6. For follow-up, tell facts about everyone who has been introduced. The class identifies the people by name and their location in the room. Alternatively, you can tell the names of Ss so that the class can tell their location and facts about them. Or have Ss stand up in turn. The class tells their names and what they remember about them from their conversations during the game.

7. Point out those Ss who have not been introduced during the game. The

class tells their names and information about them. For follow-up on another day, call roll. Ss answer *not* for themselves, but for the Ss near them (in front of them, to their right, etc.). They tell what they remember about that classmate.

LEVELS = BEGINNING TO HIGH INTERMEDIATE
↙ SUGGESTIONS FOR ADAPTATION ↘

⇓ For the lowest level, limit the number of questions to one or two, perhaps simply "What's your name?" and "How do you spell it?" Help Ss distinguish between first and last names. Teach the names of the letters of the alphabet so that Ss can spell their own names aloud and write names as they hear the letters.

⇓ For high beginners, add only simple questions such as "Where are you from?" and "What's your job?" Review possible answers and the names of the letters of the alphabet.

⇑ For higher levels, let Ss suggest possible questions and choose those most essential to the type of information being gathered. Especially if some Ss already know one another, encourage "creative" questions, such as "What is something unusual about you?" or "What do you like most about your work?"

POSSIBLE VARIATIONS

⇔ For writing and spelling (copying) practice, each S writes information about himself/herself in a square on another player's Bingo card, having that player reciprocate by writing on his/her card. Then s/he writes the

OTHER AREAS OF APPLICATION: Other kinds of personal information, such as details of Ss' jobs (e.g., *Where do you work? What are your job duties?*), interests, or family. Also suitable are topics involving personal experience and opinions--travel, health, favorite things, values, politics, etc.

As much as possible, students need to apply the language principles and rules they are learning to "real-life" situations. One way to bring "the outside world" into the classroom is to provide "realia"-- authentic "language materials" that are part of everyday life in the community.

⇒ **SPECIFIC CONTENT OF THESE INSTRUCTIONS: Restaurant conversations.**

⇒ **MATERIALS: For each group, a set of identical menus from a local eating place, preferably with illustrations of dishes available, at least half as many menus as number of Ss; writing paper.**

Parts of a sample menu from a local eating place.

INSTRUCTIONS

1. If necessary, on the board provide sample dialog or conversation frames for typical restaurant situations. Here are examples:

Host(ess):	Good evening. How many in your party?
Customer:	. . . . , please.
Host(ess):	Smoking or non-smoking?
Customer:	. . . .
Waiter/Waitress:	May I take your order?
Customer 1:	Uh...not quite yet. I need a little more time.
Customer 2:	I'm ready. Let's see...I'll have . . . .
Waiter/Waitress:	That comes with soup or salad.
Customer 2:	Salad, please.
Waiter/Waitress:	What kind of dressing would you like on that?
Customer 2:	What kinds do you have?
Waiter/Waitress:	We have . . . . , ... , and . . . .
Customer 2:	I'm sorry. I didn't get all that. Could you repeat the choices?

2. Practice these and other conversations (about choices of soup, drinks, desserts; asking for the check, etc.) with the class, having Ss supply typical phrases for the blanks. If the sets of available menus are from similar types of restaurants, you can introduce words for foods that appear often, showing pictures if available.

3. Divide the class into small groups. Distribute one set of menus to each group. Ss take roles: perhaps a host or hostess, a waiter or waitress, and a pair or group of customers. They practice a series of conversations that would be typical for the restaurant whose menus they have. Circulate around the classroom to give necessary help. If you ask them to write their conversations to hand in later, they will probably ask more specific questions about sentence structure, vocabulary, etc.

4. S groups present their conversations to the class in turn; after each "performance," ask comprehension questions (*Examples: What section did they sit in? What did Faradee order? What kinds of soup were available? Did they have dessert? How much was the bill?*). In addition, elicit comments from the class--corrections, suggestions for improvement, and questions; if there are many comments, have the group (or another group) perform it again.

LEVELS = HIGH BEGINNING TO LOW INTERMEDIATE
↙ SUGGESTIONS FOR ADAPTATION ↘

⇓ On large index cards, provide lower-level groups complete sentence frames to work from--one for each brief conversation. Ss use these to create their oral and/or written conversations.

⇑ Encourage more advanced groups to ad-lib, adding situations from their own experience (e.g., receiving the wrong item, realizing that a dish contained a 'forbidden food' that they couldn't eat, suspecting that the bill contained an error). Use the conversations the groups create to stimulate discussion about real-life situations that Ss have encountered or might encounter.

POSSIBLE VARIATIONS

⇔ Instead of having S groups create conversations for any situations that come to mind, specify the situation. (e.g., *A couple is on a blind date; they are unsure about who is going to pay for the meal*), perhaps a different one for each group. Based on the realia (the menu), groups create conversations to perform; the class describes the situation and helps them to solve the problem, if there is one.

There are many places in the community to collect realia for classroom conversations and discussions.

OTHER AREAS OF APPLICATION: Conversations based on store ads ("What can we buy?"), TV schedules ("What do you want to watch tonight?"), instruction manuals ("How does it work?" or "How do we put it together?"), bills (I think there's a mistake on my telephone bill."), bank and post office forms, and many other kinds of realia.

DYAD NARRATIVES

Sequencing events is a vital part of narration, and oral story telling is an essential language skill. Use this activity to help your students communicate and understand a joke or the point of a story.

⇒ **SPECIFIC CONTENT OF THESE INSTRUCTIONS: Cartoon strip stories without speech balloons (humor).**

⇒ **MATERIALS: A variety of picture strip stories of the same type, two copies each, one copy of which is cut apart. The panels of each story should be clipped together or kept in an envelope.**

Teaching Tips

Many reproducible ESL books are good sources of this material. Some examples are *Action English Pictures* (Alta Books) and *Comics and Conversation* (JAG Publications, Studio City, CA). Carefully-chosen newspaper comic strips work well, too. If the frames are pasted up on index cards and the intact story on card stock (laminated if possible), you are more likely to get complete sets of materials returned to you.

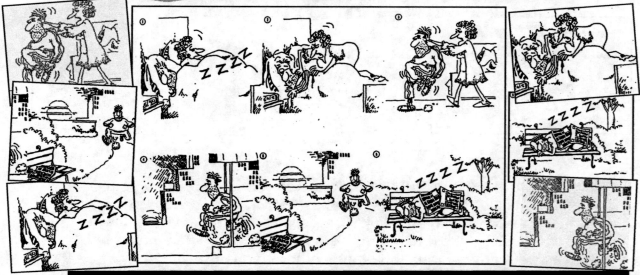

A sample "Dyad Narrative" game created from a page of *Comics and Conversation.*

INSTRUCTIONS

1. Distribute one set of strip stories to each pair of Ss; one S receives the intact story (one page), while the other is to work with the individual panels.

2. The S with the intact story narrates the events to his/her partner; without looking at the page, the partner arranges the panels in correct order, asking questions for clarification if necessary. S/he compares his/her strip story to the original.

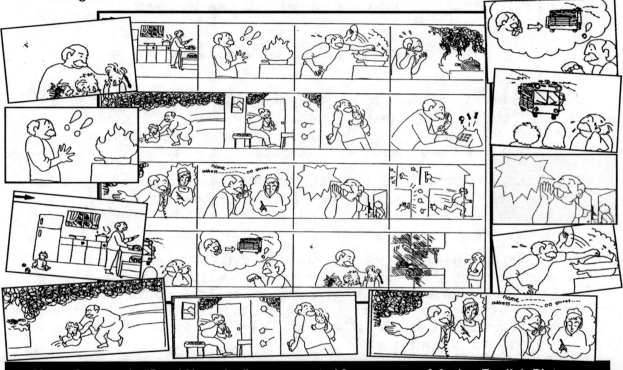

Most of a sample "Dyad Narrative" game created from a page of *Action English Pictures.*

3. Ss in pairs write the events or steps of the story in a paragraph or a numbered list (one number per panel). If the story is a joke, they write the point. Give necessary help. With both names on it, their paper may be turned in for comment, correction, and/or a grade.

4. When they finish one story, Ss exchange the materials for another set, completing as many stories as they can within the given time period.

5. After the class reconvenes, S pairs in turn tell one of their stories to the class, perhaps with one S pantomiming the action; check for class comprehension by

   • having the class take notes on the events or steps and then retell the story from their notes or

   • projecting the picture story on the overhead and having another S or Ss retell the story again, with the class making corrections.

LEVELS = HIGH BEGINNING TO HIGH INTERMEDIATE
◤ SUGGESTIONS FOR ADAPTATION ↘

⇓   For lower-level groups, write the relevant vocabulary from all the stories on the chalkboard and refer to it when Ss need help. Or you can distribute a vocabulary list of words in numbered groups, corresponding to the numbers of the stories.

⇓   Have one or more sets of "answer keys" (the stories as *you* might tell them) available for Ss to compare with their own narratives.

POSSIBLE VARIATIONS

⇔   Have Ss work in small groups; supply only one S with the picture story in its intact form. The others all receive "scrambled panels" in envelopes. The S with the intact story tells it to the group. Listeners ask questions and compare their sequencing with their classmates' versions.

⇔   Supply no picture stories in their original form; after pairs or groups of Ss have worked together to arrange the frames of the story they have received in logical order, they check their work with the "answer key" (the intact story).

⇔   During another class period, distribute the written narratives to Ss (or S pairs) other than their writers. Spread out the intact picture strip stories on a table. Ss match the narratives they have received to the picture stories. They correct the papers, if necessary, and then rewrite them in another tense form. For example, If the original is in the historical present, the receivers can rewrite it in the past.

⇔   Instead of picture strip stories with a clear sequence of events, provide pictures of actions in which the sequencing can vary. In this case, both Ss of each pair receive the same panels in envelopes. One S arranges his/her panels in any logical order and tells his/her partner the story; the partner arranges the panels in the same order. Then they reverse roles and repeat the activity, later writing the narrative in an agreed-upon sequence. As a follow-up, they can exchange their papers and panels in envelopes with another pair of Ss, who read the narrative aloud and arrange the panels according to what they have read.

Here are the missing panels from the *Action English Pictures* story on the previous page.

Because the ability to work effectively in "teams" is an important lifeskill, cooperative learning tasks are often included in workplace and academic skills training. This entertaining and engaging activity incorporates oral language practice with collaborative learning. It also involves outcomes.

OK COOPERATIVE JIGSAW PUZZLES

⇒ **SPECIFIC CONTENT OF THESE INSTRUCTIONS: Scene description.**

⇒ **MATERIALS: Commercial or self-made jigsaw puzzles of about 40 to 100 pieces each—perhaps a children's puzzle, preferably of a situation that will elicit useful vocabulary; writing paper.**

CAN YOU FIND:
the birthday candle awning, the wristwatch scale, the pretzel stick fence and over 85 other common objects that deceive the eye?

This is a jigsaw puzzle with lots of "hidden vocabulary" in it.

INSTRUCTIONS

1. Provide the original picture (on the puzzle box) and/or one or more photocopies. The group places the picture(s) in a position that all members can see. Each S begins with a different corner or straight edge piece; Ss divide the other puzzle pieces equally. At the top of a piece of paper, they write all their names and the time at which they are going to begin work.

2. Each S in turn asks the group for an adjoining puzzle piece to what s/he has thus far constructed by describing what s/he has and needs (*e.g. I have the upper right-hand corner with a mouse in front of a tree trunk. I need a piece with a straight edge on the right. It has more of the tree trunk on it.*) The S with the requested piece locates and supplies it as quickly as possible. Ss continue asking for pieces until they have all finished their section. Then they put their parts together to form the completed picture.

This part of a children's jigsaw puzzle called *Pops-Town* shows many actions to describe.

3. On their paper, the group writes the time at which they finished the puzzle and the total time (number of minutes) needed. Collectively, they compose a written description of the picture--in as much detail as possible. They also write answers to the following questions about cooperative learning (listed on an index card or on the chalkboard.)

   - What techniques helped you to complete the puzzle quickly?

   - What practices slowed you down?

   - Did you all participate equally? If so, how? If not, why not?

   - How can you work better and more quickly on the next puzzle?

4. Ss hand in group and/or individual papers for comment.

LEVELS = HIGH BEGINNING TO LOW ADVANCED
⬐ SUGGESTIONS FOR ADAPTATION ⬎

⇓ Lower-level Ss should work with easier puzzles (puzzles with larger pieces). On an index card or paper accompanying the puzzle, supply necessary vocabulary, including words for colors and shapes, the names of items in the picture, prepositions of location, etc.

⇓ To help Ss with puzzle strategy, draw lines on the photocopy of the complete picture, dividing it into quadrants or smaller rectangles; Ss will be able to relate puzzle pieces to specific sections of the picture.

⇑ After the first puzzle round, advanced Ss can work from the picture descriptions that other groups have written, instead of from the picture itself.

POSSIBLE VARIATIONS

⇔ Instead of having Ss distribute the puzzle pieces before they begin, they can spread these out on a table for all Ss to look at simultaneously. Each S in turn must locate and place a piece into the puzzle, describing the details that make it fit. To introduce an element of competition, one S can act as the timer. Ss receive a point for each piece they place only when they find it within a given time limit, such as fifteen seconds. Alternatively, all Ss can look for adjoining pieces at the same time, receiving one point for each piece they place.

⇔ Instead of one puzzle, provide two reasonably simple ones. Mix up the pieces. Ss form two teams, each team receiving a different picture. Each S in turn takes one of the puzzle pieces, telling what s/he sees in it to make him/her believe that it is part of his/her team's puzzle. When all the pieces have been claimed, the teams work on their puzzles separately. After they have completed their puzzle and described the picture in writing, they mix up the pieces and begin again, trading pictures (or picture descriptions).

OTHER AREAS OF APPLICATION: Geography jigsaw puzzles, such as those in which each piece is a separate state or country (Ss practice place names and statements of location, such as "Nevada is east of California and south of Oregon and Idaho."); homemade jigsaw puzzles, perhaps of scenes with which Ss are already familiar, mounted on cardboard and cut apart (in irregular or regular pieces, perhaps squares); for vocational English, homemade jigsaw puzzles of complex work machines, etc. (Students have to say the names of parts to complete the puzzle).

READING KIT

Beyond the very lowest level, all reading is really reading for meaning. To provide essential additional practice in understanding main ideas and supporting details, you can prepare several of these versatile and efficient "reading kits," at little or no cost.

⇒ **SPECIFIC CONTENT OF THESE INSTRUCTIONS: Narratives: news stories, anecdotes, or fables.**

⇒ **MATERIALS: A commercial or self-prepared "reading kit," consisting of reading selections of similar or somewhat varied lengths and difficulty; paper to write answers on; separate answer keys (one answer card for each reading selection or several copies of a complete answer key).**

Teaching Tips

You can easily create an appropriate reading kit by cutting up one or two copies of a reading textbook or storybook that you are no longer using; paste up each reading (with exercises, if you wish) on a separate sheet of paper or large card.

STORY 1: A Dangerous Dinner

When a man came home from work, he found a small bottle of yellow liquid in his mailbox along with his bills, letters, and advertising mail. It was a free sample. There were two lemons on the label with the words, "with Real Lemon Juice." Happily, the man added it to the salad dressing that he made for his dinner that evening.

Soon he got sick. As it turned out, the advertising sample was a bottle of dishwashing soap that *contained* lemon juice, not a food product to be used in cooking. A lot of other people got sick too--after they had put the liquid on fish, in salads, and in tea. Some people went to the hospital. Luckily, no one died from the misunderstanding.

The company that had distributed the soap soon changed its label. And a lot of customers learned to read labels more carefully--especially the labels on free samples.

STORY 2: Good Friends and Relatives

Because of his stomach cancer, Manuel Garcia went to the hospital for chemotherapy. After a few weeks of this strong treatment, Manuel's hair began to fall out. Soon he was completely bald, ashamed, and very depressed. He didn't want anybody to see him that way.

When his brother and a few other relatives came to visit him in the hospital, Manuel was shocked. None of them had any hair, either! They had shaved their heads. Manuel and the other men began to laugh. The nurse told them to be quiet, but she was smiling too.

As soon as Manuel came home from the hospital, other friends and relatives came to visit. They all asked him to shave *their* heads, too, and he did so--up to fifty heads in a day. But he drew the line when his wife wanted to be bald too!

"With such good friends and relatives," says Manuel, "now I'm ready for anything."

Some sample *Reading Kit* selections pasted up on cards (news stories).

INSTRUCTIONS

1 Explain the purposes of the kit, perhaps completing a sample selection with the class on the overhead projector or through a handout.

2 Distribute a different selection to each S. S/he reads the material. Noting the number or title of the selection, s/he writes the answers to the questions on a separate piece of paper.

Teaching Tips

If the selection itself does not include any (or sufficient) comprehension exercises, write questions such as these on the board: *What happened in the story? What is the point (the joke? the moral? the meaning?) of the story?*

3. Ss work at their own pace, using the available answer key to check their answers after completing each selection or several selections. Exchanging cards with one another or with the materials remaining in the kit, they read as many selections as they can within the given time limit. At the end of the activity, you may wish to collect their answer sheets for comment and/or to record grades.

4. Ss in turn tell the class one interesting or valuable thing they learned from the reading selections they completed. No S may repeat information already presented by another S. Orally and informally, check that the class has understood the points presented.

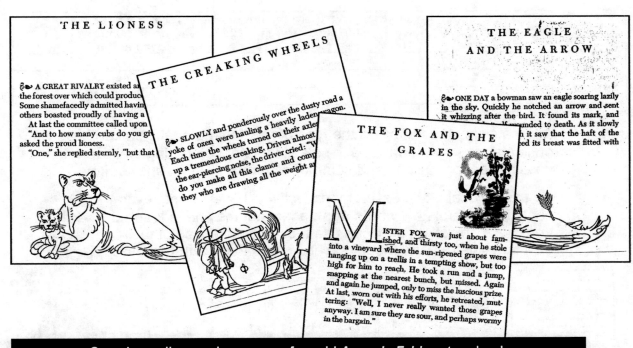

Sample reading cards—pages of an old *Aesop's Fables* story book.

LEVELS = HIGH BEGINNING TO ADVANCED
⬋ SUGGESTIONS FOR ADAPTATION ⬂

⇓ Help Ss, especially less able ones, by suggesting the readings that they should begin with and by monitoring their progress.

⇓ If answers are short, you can distribute one complete answer key to each table or group of Ss, so that they do not have to get up to check their work. They can pass each reading selection to another S in their group as soon as they finish with it.

⇔ If answers are long (essay-type, as in a summary), prepare an answer key of possible responses yourself. Make a few copies of these available to Ss so that they can compare their answers to yours. Help Ss individually by discussing their answers with them

POSSIBLE VARIATIONS

⇔ Instead of narratives (news articles, fiction stories, jokes, anecdotes, fables, etc.), create *Reading Kits* consisting of excerpts from non-fiction articles, academic readings of different types (description, explanation, definition, process, comparison, persuasion, etc.), vocational or technical reading, and/or other reading material appropriate to the level and goals of the course. In addition to (or instead of) comprehension exercises, have Ss answer these questions about each selection, perhaps in chart form:

Number or Title of the Selection	What is the main idea of the reading selection?	What are the important supporting details?	What are your comments on or opinions of the ideas in the reading?

⇔ Alternatively, depending on the type of reading material, you can have Ss outline and/or summarize each selection, later comparing their work to the possibilities for outlines and summaries provided in an answer key.

⇔ To add a vocabulary lesson to this activity, have Ss list several especially useful new words or phrases that they have learned from each selection; they can use these in sentences of their own that illustrate the meanings and share them with the class.

At no or low cost, you can increase your students' access to information through Reading Kits.

Most of the everyday reading that people do is really scanning—searching for specific information from within everyday sources such as newspapers, telephone books, signs, and the like. This competitive activity works with any appropriate materials that you have collected.

⇒ **SPECIFIC CONTENT OF THESE INSTRUCTIONS: Information from the local telephone book.**

⇒ **MATERIALS: Pages from the local telephone book with useful information, such as how to make a long-distance call, day and evening rates, annual calendar of community events, recreation facilities and attractions, theater seating, community services, First Aid instructions, bus routes, etc. For each group of Ss, one page of questions to answer—or these can be written on the chalkboard or projected onto a screen with the overhead.**

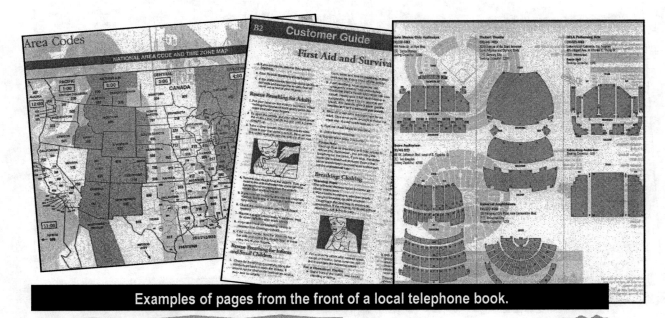

Examples of pages from the front of a local telephone book.

Doing Without the Photocopier

INSTRUCTIONS

1. Post the various pages of information (numbered) with tape or push pins on walls and boards around the classroom. Or simply place them on separate desks or tables. These are the "stations" of an "Information Scavenger Hunt."

2. Give each team of Ss a list of questions to answer or items to find. These should be divided into numbered groups, with no more than three or four items corresponding to each page of information at a station.

Teaching Tips

To prevent teams from "bunching up" at the same stations, you might want to arrange the groups of questions or items given to each team in a different order; if you wish to reuse these papers, instruct teams not to write on them.

3. Understanding that there is a time limit, Ss in teams circulate around the classroom, cooperating with team members in finding the requested information as quickly as possible. They write the answers on the handout or a separate piece of paper. The first (or only) group to finish within the established time limit gets a small prize.

4. Reconvene the class and go over the answers; teams correct their own papers. Their total score is the number of items correctly completed. The team with the highest score receives a prize that can be shared among its members.

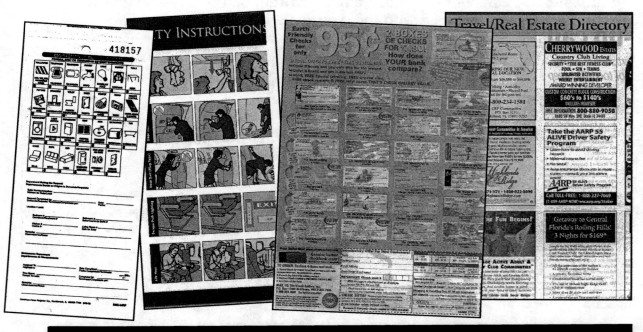

Examples of other kinds of realia suitable for "Information Scavenger Hunts."

LEVELS = HIGH BEGINNING TO LOW ADVANCED
⬋ SUGGESTIONS FOR ADAPTATION ⬊

⇓ Adapt the level of the realia and of the questions, as well as the number of stations and the number of questions, to the language level of the Ss. For instance, lower-level Ss might be asked to find only that information that is in bold print or in numbers.

⇓ To make the game easier, present the essential vocabulary before beginning the activity. Encourage higher-level Ss to jot down apparently important words that they encounter during the game and to write these on the board. Go over the items after the game.

POSSIBLE VARIATIONS

⇔ If space is limited, instead of having Ss move around the room in groups, pass the materials around the classroom. Or if there are enough materials (such as local telephone books) available for each S or pair of Ss, distribute these to the groups. After explaining that the "winner" of the game is the team to finish first, you may or may not want to present a brief lesson on techniques of teamwork. Alternatively, each team can figure out its own methods of working together, and these can be the subject of a follow-up discussion.

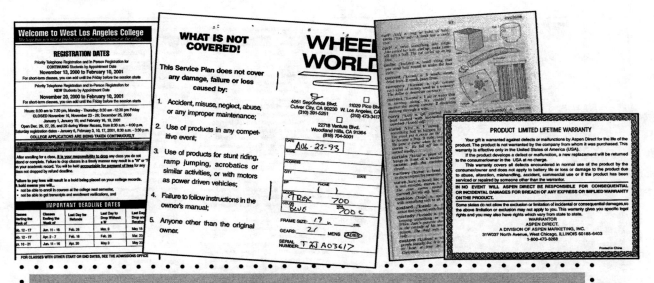

OTHER AREAS OF APPLICATION: Any realia of interest and use to Ss, such as ads, transportation schedules, catalogs, brochures, dictionary pages, charts, maps, product cards, and many others.

READING AUTHENTIC MATERIAL

All language students living in an English-speaking culture need to "understand" material written on a level beyond that at which they are comfortable reading. Here is a basic training activity that helps students comprehend authentic reading materials.

⇒ **SPECIFIC CONTENT OF THESE INSTRUCTIONS:** Campus brochure on "Personal Safety Tips to Reduce Crime."

⇒ **MATERIALS: A class set of authentic materials that can be obtained at little or no cost, such as fliers or brochures from your school, a local bank, the post office, medical clinics, etc.**

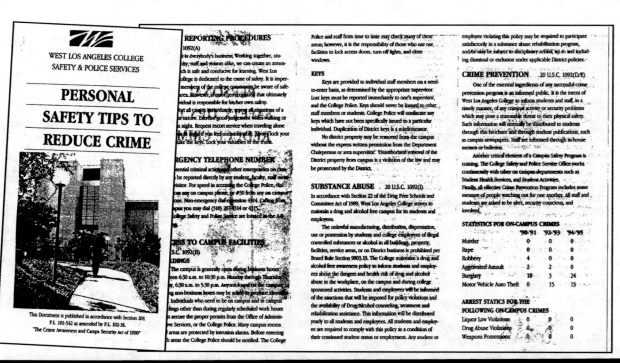

An example of authentic reading material useful to students at various levels of language ability.

INSTRUCTIONS

1. Distribute the realia to Ss in groups; as a pre-reading exercise, ask general questions about it (e.g., What is it? What is it for? Where is it from? What are the main parts?)

2. Distribute a question list to each group or write questions about the main ideas of the realia on the board. *Examples*:

    a. What two kinds of crime does the brochure cover?

    b. How can you contact the campus police?

    c. To reduce crime, how and where should you walk?

3. Ss work together as fast as they can to find the answers in the reading material. They compile one answer list.

4. Keeping Ss in groups, go over the answers with the class. Give a prize to (a) the first group to finish, and (b) the group with the most correct answers.

5. If desired, use this activity as a lesson on group interaction: find out the techniques that each group used in order to finish the task quickly; discuss the effectiveness of each method and ways to improve speed and accuracy through cooperation.

Other examples of free (public service) realia for classroom use.

LEVELS = INTERMEDIATE TO ADVANCED
↙ SUGGESTIONS FOR ADAPTATION ↘

⇓ For lower-level groups, write important vocabulary on the board and discuss it before beginning the activity. Present or review scanning techniques, such as moving the eyes down the middle of the page, looking for numbers or capital letters, etc.

⇓ For lower-level and mid-level groups, provide question types other than *wh*-questions (i.e., questions beginning with *what, why, how,* etc.)

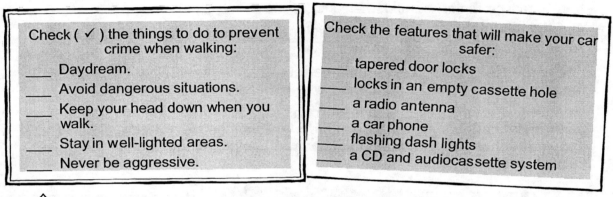

Check (✓) the things to do to prevent crime when walking:

____ Daydream.
____ Avoid dangerous situations.
____ Keep your head down when you walk.
____ Stay in well-lighted areas.
____ Never be aggressive.

Check the features that will make your car safer:

____ tapered door locks
____ locks in an empty cassette hole
____ a radio antenna
____ a car phone
____ flashing dash lights
____ a CD and audiocassette system

⇑ Higher-level groups can write their own list of questions, exchanging it with that of another group to complete.

POSSIBLE VARIATIONS

⇔ Ss compete individually to find the answers as quickly as possible and to write accurate answers.

You can find free or low-cost "realia" in many places—at home and in the community.

OTHER AREAS OF APPLICATION: brochures, such as from banks, medical clinics, the post office, and so on; government booklets on various topics; throwaway advertising; free local papers; free magazines with job ads; etc.

N. Reading Authentic Material

In academic, creative, and communicative writing, every sentence must have meaning. To demonstrate the importance of a single sentence and the many ways an idea can be expressed, try this sentence-composing activity.

⇒ **SPECIFIC CONTENT OF THESE INSTRUCTIONS: Pictures of events (simple sentence structure).**

⇒ **MATERIALS: Pictures of situations that can be explained in one sentence each, such as illustrations, original photographs, photos from newspapers or magazines, picture postcards, book illustrations, or cartoons; writing paper.**

These examples of one-frame illustrations from *A Journey Through America* show stages in U.S. immigration history. They can be captioned in one sentence each.

INSTRUCTIONS

1. If necessary, review basic sentence structure: *subject + predicate*. Point out the differences between complete sentences and sentence fragments, run-ons, comma splices, and the like. Using sample pictures, have the class discuss (and agree on) what is going on; help Ss compose *one* sentence for each picture that explains the situation. In other words, teach caption writing: emphasize that an effective caption must be (a) accurate, (b) concise, and (c) complete. Here are some *examples of possible captions for some of the pictures on the previous page:*

   - In the 15th century, there were Indians ("native Americans") living on the North and South American continents.
   - Many of the early European immigrants were fishermen, traders, and farmers.
   - The Spanish built Spanish-style houses in the Southwest.
   - Black Africans were captured and sold by slave traders and forced to work on plantations.
   - After the Gold Rush in California, poor Chinese came to work in mining camps.

2. Divide the class into small groups; divide consecutively-numbered pictures into equal sets; give one set to each group. Ss in groups discuss what is going on in each picture and compile a list of captions--one sentence per picture. These should be written on the left side of a piece of paper.

3. Groups exchange papers and picture sets. One S shows each picture to the group; others decide on and read aloud the matching sentence from the first group's paper. Discussing the situation, the group writes an alternative or an improved caption on the right side of the paper.

4. Collect each set of pictures and the corresponding paper; show the picture and read aloud the two captions. The class decides which is the better sentence and tells why.

Instead of removing the captions from one-frame cartoons, you can blot out the words in the speech balloons of cartoon strips. See the *Possible Variations* on the next page.

LEVELS = HIGH BEGINNING TO LOW ADVANCED
⬋ SUGGESTIONS FOR ADAPTATION ⬊

⇓ For lower levels, provide a handout on basic sentence patterns, using possible picture captions as examples. As an exercise, Ss identify the sentence parts--e.g., subject, predicate, time expression, place expression, etc. Use these examples to teach the elements of a concise, complete, correct sentence.

⇑ For higher levels, teach or review the structure of compound sentences, providing lists of connecting words (*and, but, so*, etc.). Repeat the activity described above with the same or different pictures (perhaps pictures with clearly-contrasting elements), having Ss produce a two-clause caption for each. Alternatively or additionally, repeat with complex sentences (independent + dependent clauses, with connecting words such as *because, when, as soon as, if, although*, etc.).

POSSIBLE VARIATIONS

⇔ Ss work on the pictures individually or in pairs, writing their own lists of captions. You may or may not wish to provide an answer key of possible captions. Display pictures on a board or table. As Ss read aloud each caption, others identify the matching picture.

⇔ Use pictures that already have captions, such as newspaper or magazine photographs, or one-frame cartoons. Before beginning the activity, remove the captions from the pictures. Paste up or retype the captions (in larger type), either on separate index cards or in a list on a piece of paper (or simply copy the captions onto the chalkboard). After Ss have created their own captions, they try to find the "real" caption. In turn, Ss show their pictures and present the two captions to their classmates, who must (a) identify the "real" and the new caption, and (b) tell which caption is better, and why.

⇔ Alternatively, use one-frame cartoons from publications in languages Ss don't know. Ss guess what the captions might mean before receiving the translations.

⇔ Instead of one-frame cartoons, present cartoon strip stories with the speech balloons removed or blotted out. Ss create their own conversations before comparing their work with the original.

„פעם הבאה תעני על הטלפונים שלך בעצמך..."

An example of a one-frame cartoon in a language that students—and the teacher—might not know.

CHAIN WRITING WITH VISUALS

"**I**n real life, the purpose of all writing is to communicate." Help your students to get across their message efficiently and effectively by presenting writing activities that provide immediate and real feedback. You and they will enjoy the many creative variations of this Chain Writing activity.

⇒ **SPECIFIC CONTENT OF THESE INSTRUCTIONS: Street directions.**

⇒ **MATERIALS: Paper to draw and write on, preferably long paper (8 1/2 x 14 ").**

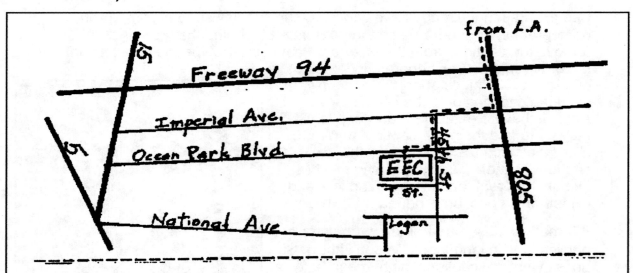

If you are coming from Los Angeles, take Freeway 805 south to the Imperial Avenue exit. Turn right onto Imperial. Turn left on 45th Street. Go to Ocean Park Boulevard (where there is a stop sign) and turn right. The EEC (Educational Cultural Complex) is on the left. There is a large parking lot directly in front of the school. Also,...

An example of a *Chain Writing* activity—on street directions—in progress.

INSTRUCTIONS

1. Ss divide their papers into equal thirds. On the first third, they draw a street map illustrating how to get from one place to another, such as from school to their home or vice versa. If distances are long, they can leave out details. You might make city maps available (in a large city, the relevant pages of the city map book) so that Ss can check the accuracy of their drawings.

2. In the second third of their paper, Ss describe their directions map in words, making sure that their description corresponds to the drawing.

3. Ss fold the first third of their paper (the map) over so that it cannot be seen; they pass the paper to a partner, who--in the last third of the paper--draws a direction map according to the words in the paragraph above.

4. Each S compares his/her map at the bottom of the paper to the original at the top. If they are not the same or adequately similar, Ss figure out when communication broke down--i.e., in the writing, reading, or drawing? They make necessary improvements and hand in their papers for comment, correction, and/or grades.

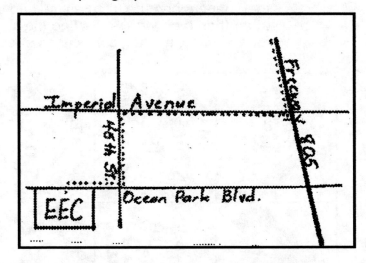

LEVELS = HIGH BEGINNING TO LOW ADVANCED

⇓ For lower-level classes, provide copies of local maps to "draw" directions on--i.e. instead of drawing a "direction map," Ss simply draw a line on an available one. They then write the directions in words; on another map, their partners draw a line according to the directions they read.

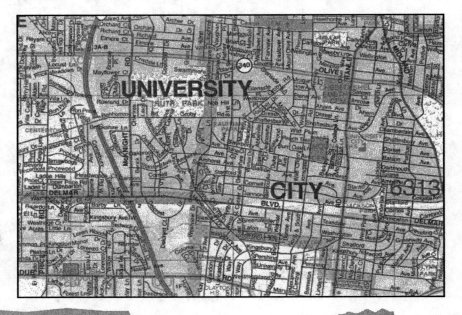

POSSIBLE VARIATIONS

⇔ Instead of having Ss draw maps or other illustrations to describe in words, you can provide sets of pictures for chain-writing activities. For example, you could distribute magazine pictures of different kinds of people (of various ages and races, clothed differently, etc.), one to each S or pair of Ss. After Ss write a description of the person in their picture, collect both their papers and the pictures. Mix up the papers and redistribute them to Ss other than their writers. Ss find the matching picture and check the paper for accuracy, making corrections or suggesting improvements if necessary..

Examples of the kinds of pictures useful for a "Chain Writing" activity. These student self-portraits are from the cover of *L.A. Mosaic*, complied by Laura Silagi.

Why not turn the humdrum filling out of forms into creative "Chain Writing" assignments? Again, the key to success is the immediate feedback that students receive on the communicative effectiveness of their writing. As you engage them in this cooperative multi-skills activity, you will also be cutting down on your correction time.

⇒ **SPECIFIC CONTENT OF THESE INSTRUCTIONS: Housing information forms.**

⇒ **MATERIALS: Two class sets of a form to fill out, such as a school enrollment form, a housing application, a credit-card application, a telephone message form, an office memo form.**

TENANT'S PERSONAL AND CREDIT INFORMATION
(In the event of co-tenants, other than spouses, use separate sheet for each tenant.)

Name	Date of Birth	Social Security No.
		Drivers Lic. No. Expir. Date
Name of Co-Tenant		Social Security No.
Present Address		Drivers Lic. No. Expir. Date
City/State/Zip	Res. Phone	Bus. Phone
How long at present address Landlord or Agent		Phone
Previous Address How long Landlord or Agent		Phone
City/State/Zip		
Occupants: Relationships:		Pets?
Ages:		
Car Make Year Model	Color	License No.

OCCUPATION

	PRESENT OCCUPATION *	PRIOR OCCUPATION *	CO-TENANT'S OCCUPATION
Occupation			
Employer			
Self-employed, d.b.a.			
Business Address			
Business Phone			
Type of Business			
Position held			
Name and Title of Superior			
How long			
Monthly Gross Income			

For each student, provide two identical copies of the same blank form.

INSTRUCTIONS

1. Each student fills out a form with information about himself/herself.

2. S/he passes the form to a classmate, who writes a paragraph that includes all the information on the form. *Example:*

> *Cindy Wu's birthday is March 17. She's 24 years old now. She lives with a roommate at 8821 Michigan Avenue in Glendale, California, 85853. She has been at this address for 2 1/2 years. Her landlord is Peter Wong. She has a social security number, but she doesn't have a driver's license. . . .*

3. The classmate passes the paragraph to a third S, who fills out an identical blank form with information from the paragraph.

4. The group compares the two forms about each S, figuring out where communication broke down, if it did, and making necessary improvements. Each S hands in two comparable forms, stapled to the paper s/he wrote, for comment, correction, and/or a grade.

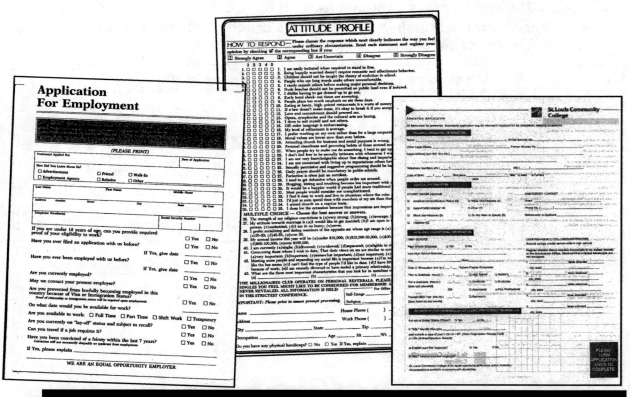

Some ideas for other kinds of forms to use for *Chain Writing*: school enrollment forms, weekly schedule forms, bank and post office forms, job application forms, computer dating forms, etc.

LEVELS = HIGH BEGINNING TO LOW ADVANCED
↙ SUGGESTIONS FOR ADAPTATION ↘

⇓ For lower-level classes, you might provide simplified forms or parts of forms. Add necessary steps, such as filling out a sample form on the overhead before having Ss fill out their own forms. Review the sentence patterns they will need for paragraph writing.

⇑ For higher-level classes, add challenge by substituting a pair interview for Step 1 above: instead of filling out a form for himself/herself, each S interviews a classmate and fills the form out for him/her. S/he then writes the paragraph about the form s/he has completed.

POSSIBLE VARIATIONS

⇔ Instead of forms, you can use charts or graphs for a "Chain Writing" activity. In this case, Ss will have nothing to fill out, so they will skip Step 1 above. Instead, each S writes a prose description of the information in the graph or chart. S/he passes his/her paper to a partner, who either creates a comparable graph or chart from the information or applies it to an identical graph or chart from which elements have been deleted (e.g., lines from a line graph, bars from a bar graph, numbers from a chart).

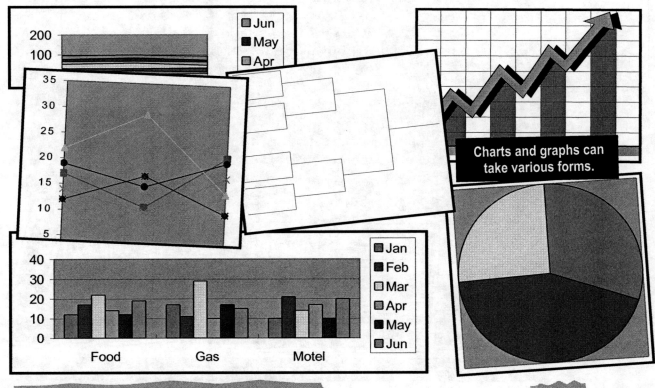

Charts and graphs can take various forms.

VOCABULARY CONCENTRATION

Most vocabulary acquisition occurs in context. On the other hand, there are many vocabulary activities that not only amuse and challenge but that also help students learn new words systematically. This word card game can be played competitively or cooperatively.

⇒ **SPECIFIC CONTENT OF THESE INSTRUCTIONS: Compound words.**

⇒ **MATERIALS: Writing paper; decks of 20 to 40 word cards each. Each deck consists of 10 to 20 pairs of cards--each card with half of a compound word on it.**

Teaching Tips

(a) The game will be easier if you print the first half of all the words (including hyphens if necessary) on index cards of one color and the second half on cards of a different color. (b) More than one word can have the same beginning (e.g. _homework_ / _homesick_ / _homeland_) or the same ending (e.g., _headache_ / _heartache_), but it is probably better for each deck of cards to consist of words that all begin with the same one or two letters. Otherwise, there may be several correct matching possibilities, and the cards will not "come out even." (c) The decks should be shuffled so that the matching cards are not next to each other and kept together with rubber bands in small envelopes. (d) Label each envelope as to contents -- e.g. 15 _compound words beginning with H / Intermediate Level._

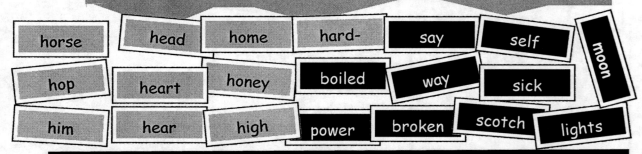

horse · head · home · hard- · say · self · moon · hop · heart · honey · boiled · way · sick · him · hear · high · power · broken · scotch · lights

A sample *Vocabulary Concentration* game, based on compound words.

INSTRUCTIONS

1. Divide the class into small groups of approximately equal size. In a multi-level class, each group could consist of Ss with similar abilities or of Ss with mixed abilities.

2. Give each group (or have each group choose) a different deck of word cards. Ss spread out the cards face up on a table or desk and work together to put them together in compounds. Encourage dictionary use.

Teaching Tips

(To save Ss time, you might include an "answer key" of correct compound words--on an index card in the envelope. Encourage group members to work together to match the cards on their own before checking their answers with the key.

3. Ss list examples that illustrate the meanings of the words they have created, underlining these words, perhaps adding definitions. One group member prepares a master list to hand in, with all group members' names at the top, perhaps in chart form with these headings:

Compound Word	Part of Speech	Definition (Meaning)	Phrase or Sentence Example That Illustrates Meaning
hopscotch	noun	a children's game	I used to play <u>hopscotch</u> as a child.
homesick	adjective		

Correct or comment on these, recording the same grade for each group member, if desired.

4. During the same time, group members list the words their groups have created on the chalkboard. If there are many words or many groups, specify the number of words that each group should contribute to the list on the board. They should choose especially useful words or words that are new to them.

5. After most of the groups have completed their tasks, reconvene the class. Teach the words on the board quickly by using these techniques: for each column of words, give word cues, such as definitions, synonyms, or sentences with "oral blanks." The class pronounces the appropriate word. After each column or after all words have been covered, give an oral quiz by having the class recall the words in response to cues without looking at the board.

6. Ss may want to copy the word lists from the board. As a follow-up during the next class period, give a written or oral quiz on the words.

LEVELS = HIGH BEGINNING TO LOW ADVANCED
↙ SUGGESTIONS FOR ADAPTATION ↘

⇔ Adapt the level of each deck of cards to Ss' language level. For lower-level groups, include only commonly-used words. For higher-level groups, include more unusual and more interesting compounds. More advanced Ss can also work with larger decks than beginners. You can mark various decks of cards with a level designation so that groups can choose the appropriate ones.

⇔ During the whole-class oral vocabulary quiz, give word cues in order if the Ss are at a low level of language proficiency. Say cues in random order if they are at a higher level.

POSSIBLE VARIATIONS

⇔ Have Ss create compound words by playing the game of *Concentration*. Instead of matching visible word parts, each group spreads out the cards in its deck face down. The first player turns two of the cards face up. If they are a match, s/he pronounces and shows the group the compound word, keeps the pair, and (perhaps) takes another turn. If the word parts do not belong together, s/he turns the cards face down again, and play passes to the next S. Ss try to remember the locations of the word parts so they can make matches more quickly. After all the cards have been collected, the winner is the S with the most cards. (You might provide small prizes.) Ss then list their words in examples and write them on the board as described.

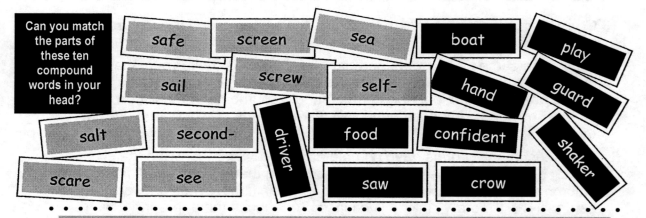

O THER AREAS OF APPLICATION: Any vocabulary areas that involve the matching of two elements, words, or parts of phrases-- such as prefixes and word roots (e.g., *pre + view*); roots and suffixes (e.g., *sad + -ness*); homophones (e.g., *haul / hall*); parts of irregular verbs (e.g., *buy / bought*); verbs and nouns that fit together in phrases (e.g., *take / a bath*); etc.

When students acquire new vocabulary systematically, it is important that they learn how to use the words and phrases in context. In order to do so, they must recognize the appropriate parts of speech and definitions. Then they'll be able to create sentences of their own with the new items.

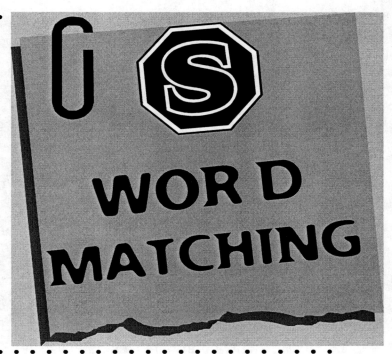

WORD MATCHING

⇒ **SPECIFIC CONTENT OF THESE INSTRUCTIONS:** Pairs of homophones.

⇒ **MATERIALS:** Writing paper; decks of 20 to 40 word cards each. Each deck consists of 10 to 20 pairs of cards--each card with a different one of a pair of homophones (words with the same pronunciation but different spellings and meanings). On the back of each card is a simple definition of the word.

Teaching Tips

(a) The game will be easier if you print one word of each pair on index cards of one color and the matching words on cards of a different color. (b) It doesn't matter what letters of the alphabet the words in each deck begin with, but if you are creating a large number of decks, it may help to systematize them in some way. (c) Shuffle the decks so the matching cards are not next to each other. Keep them together with rubber bands in small envelopes. (d) Label each envelope as to contents --e.g. *DECK 1: 15 Pairs of Homophones / Intermediate Level.*

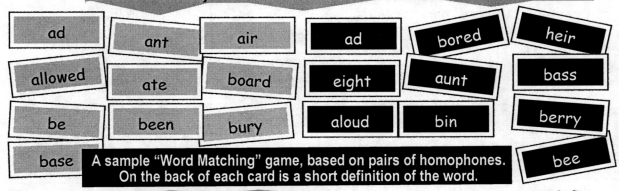

ad	ant	air	ad	bored	heir
allowed	ate	board	eight	aunt	bass
be	been	bury	aloud	bin	berry
base					bee

A sample "Word Matching" game, based on pairs of homophones. On the back of each card is a short definition of the word.

INSTRUCTIONS

1. Divide the class into small groups of approximately equal size. In a multi-level class, each group could consist of Ss with similar abilities or of Ss with mixed abilities.

2. Give each group a different deck of word cards. Ss spread out the cards face down on a table or desk and work together to match them from the definitions on the back. Encourage dictionary use.

3. Ss list sentences that illustrate the meanings of the pairs of words, underlining the homophones. One group member prepares a master list to hand in, with all group members' names at the top. Correct or comment on these, recording the same grade for each S if desired.

to exist

to put numbers together, as in math

what we need to breathe

a relative: your uncle's wife

consumed (past tense

a stinging insect that makes honey

a flat piece of wood for building

person that inherits money when someone

short word for "advertisement

uninterested; tired of a subject or activity

a small crawling insect

a number: four plus four

4. During the same time, group members list the pairs of homophones from their deck on the chalkboard. If there are many words and many groups, specify the number of words that each group should contribute to the list on the board: they should choose words they consider especially useful and/or words that are new to them.

5. After most of the groups have completed their tasks, reconvene the class. Teach the words on the board quickly by using these techniques: for each column of words, give pairs of word cues, such as definitions, synonyms, or sentences with "oral blanks." The class pronounces the appropriate pair of homophones. After each column or after all words have been covered, give an oral review quiz by having the class recall the words in response to cues without looking at the board.

6. Ss may want to copy the lists of word pairs from the board. As a follow-up during the next class period, give a written or oral quiz on the words.

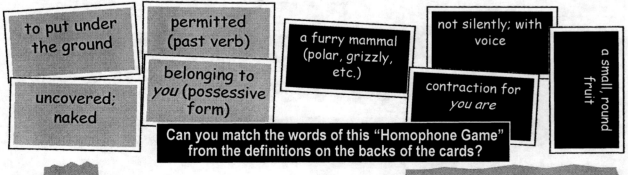

to put under the ground

permitted (past verb)

a furry mammal (polar, grizzly, etc.)

not silently; with voice

a small, round fruit

uncovered; naked

belonging to *you* (possessive form)

contraction for *you are*

Can you match the words of this "Homophone Game" from the definitions on the backs of the cards?

S. Word Matching

LEVELS = HIGH BEGINNING TO LOW ADVANCED
↙ SUGGESTIONS FOR ADAPTATION ↘

⇓ Adapt the level of each deck of cards to Ss' language level: for lower-level groups, use only common words; for higher-level groups, include more interesting words. More advanced Ss can work with larger decks than beginners. You can mark various decks of cards with a level designation so that groups can choose the appropriate ones.

⇓ On a card that is larger or of a different color card from the word cards, include a list of the pairs of words in the deck as an answer key. Lower-level Ss may need to study the words *before* playing the matching game.

⇑ For advanced groups, create decks of cards in which three words match as homophones (e.g., *pair / pare / pear; too / to / two, there / they're / their*).

⇔ During the whole-class "oral vocabulary quiz," give word cues in order if the Ss are at a low level of proficiency. Say cues in random order if the Ss are at a higher level.

***O**THER AREAS OF APPLICATION: Any vocabulary areas that involve the matching of two or more words--such as homophones (e.g., haul / hall); homographs or heteronyms (e.g., console [noun] / console [verb]); or easily-confused words (e.g., biography / bibliography).*

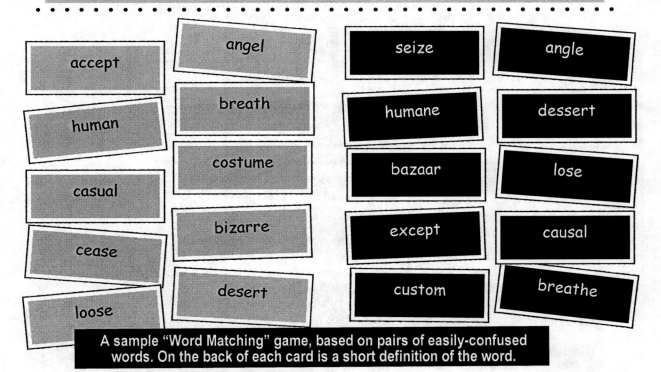

accept · angel · seize · angle · human · breath · humane · dessert · casual · costume · bazaar · lose · cease · bizarre · except · causal · loose · desert · custom · breathe

A sample "Word Matching" game, based on pairs of easily-confused words. On the back of each card is a short definition of the word.

VOCABULARY PICTURE CHAIN

Students tend to pick up vocabulary at their own rate and at their own level--as they need it to express what they want to say. Use this picture activity as a catalyst for vocabulary learning in one or more subject areas.

⇒ **SPECIFIC CONTENT OF THESE INSTRUCTIONS: Everyday objects.**

⇒ **MATERIALS: A set of pasted-up pictures of items in a category, numbered sequentially--at least one picture per S; perhaps a corresponding numbered answer key.**

 Teaching Tips

If you choose to present a vocabulary lesson on the names of small objects, you can use the items themselves instead of pictures.

   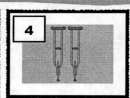 

These sample pictures of everyday objects are computer clip art, but you can find and paste up similar pictures from children's books, picture dictionaries, magazines, and many other sources. It is best, of course, to choose pictures of items that correspond to students' interests and language needs. Pictures can all belong to the same category (e.g., food, furniture, tools, clothing, etc.) or to a variety of categories.

INSTRUCTIONS

1. Arrange the chairs of the entire class into one large circle or divide the class into small groups with chairs in circles. alternatively, if desks are immovable, "cluster" Ss together, making sure there are no empty seats in the rows of desks.

2. Demonstrate the point of the vocabulary chain by showing a picture and talking about it briefly at Ss' language level. *Examples:*

   - <u>Beginning</u>: *These are sunglasses.*

   - <u>High Beginning</u>: *These are sunglasses. You wear them on sunny days.*

   - <u>Intermediate</u>: *People wear sunglasses on sunny days. They protect your eyes from the bright sun.*

   - <u>High Intermediate</u>: *People wear sunglasses for protection from the bright sun. Good sunglasses block out the ultraviolet rays. You can get prescription lenses if you need them.*

3. Then hand the picture to a S and have him/her repeat or paraphrase what you said to the class. S/he hands the picture to another S, who repeats or paraphrases what s/he just heard, and so on until the class gets the idea. Repeat with a few more pictures.

4. Hand one picture to one S in the circle(s) and talk about it briefly. That S shows the picture to the S on his/her right and repeats or paraphrases what s/he just heard, and so on around the circle. While the first two Ss are busy, immediately begin the vocabulary chain with another picture at another point in the circle. Continue introducing pictures quickly into the circle until all Ss are busy either talking or listening.

Teaching Tips

You may have to help move the pictures around by redistributing those that have piled up on one S's desk to Ss that have no pictures; you can also continue introducing new pictures into the circle. (If the desks are arranged in rows, pictures should be moving down one row and up the next; you may need to take pictures from the last S in the room and bring them to the first.)

Teaching Tips

Don't worry if Ss forget the original words you told them and are talking about the pictures in their own words. They will be acquiring vocabulary at their own level and helping one another. They will also be motivated to ask you for words they have forgotten or didn't understand.

5. While Ss are talking, give individual help. On the chalkboard, you can list vocabulary items they ask about. Alternatively or additionally, you can provide an answer key--a list of items (groups of words, phrases, or sentences) numbered to correspond to the pictures.

6. If you have listed the names of the pictures on the board, go over these words with the class. Give cues for the items in random order and have the class locate and pronounce them. Collect the pictures by asking for them by number, by name (e.g., *Who has the sunglasses?*) or by category (e.g.*, Please give me all the pictures of tools.*). For review, show each picture as you receive it and have the class talk about it.

7. Follow up with activities that include the vocabulary and in which you can actually use the pictures--for example, for food pictures, restaurant or supermarket lessons.

LEVELS = HIGH BEGINNING TO LOW ADVANCED
↙ SUGGESTIONS FOR ADAPTATION ↘

⇓ For lower levels, talk about each picture with the whole class, listing necessary vocabulary on the board, before beginning the vocabulary chain.

⇑ At higher levels, encourage Ss to add their own sentences and ideas, teaching one another new vocabulary.

⇔ Instead of simply talking about the pictures, Ss—individually and in pairs, can write down a predetermined number of words, phrases, or sentences about each picture in a numbered list. You can specify the sentence patterns to be used by listing questions on the board. (e.g., *What is this object? Where can you get it? What is it made of? How do people use it?*) Supply a numbered answer key of possible reponses.

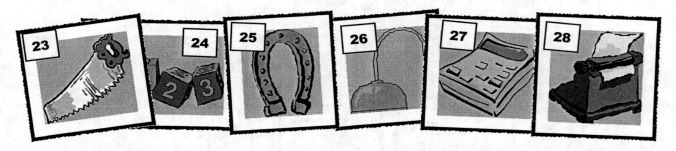

OTHER AREAS OF APPLICATION: Any content area for which a large number of different pictures are available, such as household items, work-related objects, people performing actions, sports, nature scenes, kinds of weather, people in jobs, etc.

Language learners who can "play with words" acquire vocabulary more quickly and easily because they enjoy the process. Increase your students' motivation through this challenging and amusing puzzle activity—and others that play with words.

⇒ **SPECIFIC CONTENT OF THESE INSTRUCTIONS: Phrase rebuses.**

⇒ **MATERIALS: Index cards with a number of different vocabulary items printed on them, such as rebuses--at least a few more items than there are students; writing paper.**

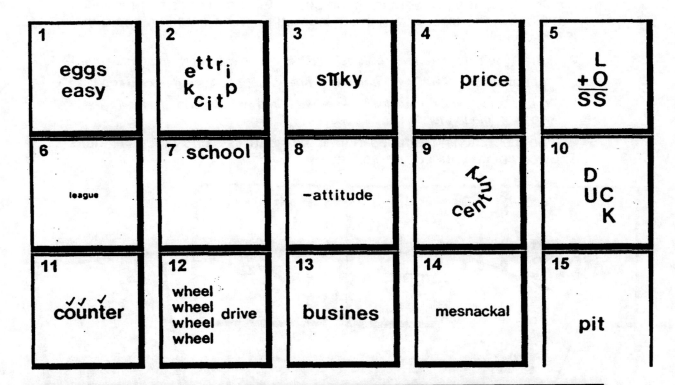

Doing Without the Photocopier

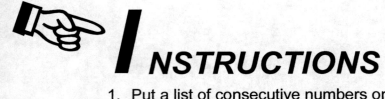

INSTRUCTIONS

1. Put a list of consecutive numbers on the chalkboard, one for each item in the "Word Play" activity, followed by space to write a phrase.

2. Demonstrate the point of the activity by "solving" a few of the items with the whole class: if the print is large, you can hold up the cards, or you can show the items on the overhead projector or copy them onto the chalkboard. Write the solutions to those items after the corresponding numbers on the board.

3. Ss work in pairs or small groups. Distribute the word play cards, making sure that each pair or group has at least one to work with at all times. Cooperatively, Ss try to figure out the solution to each item, listing their guesses in a numbered list on a piece of paper. As they finish with each card, they pass it to the next pair or group.

4. At any time during the activity, Ss who feel confident in their answers can copy them onto the chalkboard next to the corresponding numbers.

5. When time is called, collect the word cards. After showing each card to the class again, check the answer on the board. Discuss the meaning of the phrase, including its cultural background and the context in which it might be used.

6. Finally, erase the answers from the board. Give oral cues by presenting situations to the class in which the learned phrases would be appropriate. *Examples:*

   • *What would you like for breakfast? I'd like ......* (Answer: *eggs easy over*)

   • *I'm traveling to San Diego and back. I need a .....* (Answer: *round-trip ticket*)

   • *Your plan is exciting but it will never work in the real world. It's really just . . .* (Answer: *pie in the sky*)

7. Ss find the matching phrases in their individual lists or try to recall them. They complete your cue sentence.

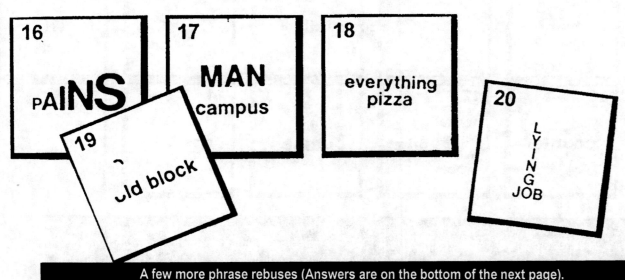

A few more phrase rebuses (Answers are on the bottom of the next page).

LEVELS = HIGH BEGINNING TO LOW ADVANCED
↙SUGGESTIONS FOR ADAPTATION↘

⇓ Adapt the difficulty of the items to the knowledge level of the class; Ss born in the United States or Ss who have been here for a long time will be able to solve rebuses more easily because they have had more cultural experience than newcomers.

⇓ If the items are too challenging for Ss to figure out on their own, you might want to supply an "answer list" of solutions in random order; Ss number the answers so that they correspond to the word play cards.

⇑ Very advanced Ss may be able to create word play rebuses of their own-- so that you can add to the items in your kit.

POSSIBLE VARIATIONS

⇔ Create and present "Word Play" items according to the steps of the "Vocabulary Concentration" activity (Idea R): on index cards of a different color, write either the solutions to the items or cues that match the solutions (e.g., *This is a way of preparing eggs for breakfast.*) Ss work together in pairs or small groups to match the cards. They list the solution phrases in sentences that illustrate their meanings.

OTHER AREAS OF APPLICATION: short traditional rebuses, personalized license plates, acronyms (e.g. *VIP = very important person; BLT = bacon, lettuce, and tomato [sandwich]; ASAP = as soon as possible*); abbreviations (e.g., *etc. = etcetera; St. = Street or Saint*); foreign words and phrases common in English (e.g. *ad hoc = with respect to this condition* [Latin]; *cul-de-sac = dead end* [French]; *salsa = hot sauce or a style of dance music* [Spanish]); idioms (the actual words); and so on.

Can you figure out these personalized license plate word puzzles?

ANSWERS TO "PERSONALIZED LICENSE PLATES" on this page: 1. So sue me. 2. Holy Cow! 3. Exquisite; 4. I'll be see in' you.

ANSWERS TO "WACKY WORDIES" on pages 61-62: 1. eggs over easy; 2. round-trip ticket; 3. pie in the sky; 4. half price; 5. total loss; 6. little league; 7. high school; 8. negative attitude; 9. turn of the century; 10. sitting duck; 11. check-out counter; 12. four-wheel drive; 13. unfinished business; 14. between-meal snack; 15. bottomless pit; 16. growing pains; 17. big man on campus; 18. pizza with everything on it; 19. chip off the old block; 20. lying down on the job.

THE
EXPERT
GAME I
(SMALL GROUP)

For both instructors and students, the most efficient and effective way to learn a subject is to teach it to others! In "the Expert Game," students become "experts" in a small amount of subject matter and take responsibility for communicating that information to others. You can use this versatile activity as part of any "content-based" lesson.

⇒ **SPECIFIC CONTENT OF THESE INSTRUCTIONS: U.S. History (Presidents).**

⇒ **MATERIALS: For each group of three to six students: a corresponding number of different short articles or stories at Ss' reading level, pasted up on large index cards; paper for taking notes and answering quiz questions.**

1.
Before the United States won independence from British rule, George Washington was a farmer in the colony of Virginia. He served as a military leader in the Revolutionary War. The colonists trusted him because he did not want power for himself. He wanted all the states and the people to work together as one. He wanted the government to serve the people well.

Washington said that power should belong to institutions, not to men. He also said that people could understand the U.S. Constitution in many ways, not just one. He did not think that the United States should have strong ties with other countries.

George Washington was the first President of the United States from 1789 to 1796. He is often called "the Father of Our Country."

2.

Thomas Jefferson could do many things. As a young man, he was a farmer and a lawyer in Virginia. He was also a scientist, an inventor, a philosopher, and an architect. He designed his own home, called Monticello. He could communicate in French, Italian, Spanish, Latin, and Greek.

Many of Jefferson's ideas became basic principles of the government of the United States. For example, he believed that "all men are created equal" (are born the same and should receive the same treatment under the law). He also said that power must come from "the consent of the governed" (the voters, not the leaders). He wanted free elections, a free press, and free speech.

Thomas Jefferson held many important government jobs. He was Ambassador to France, Secretary of State (under George Washington), Vice President (under John Adams), and the third President of the United States, from 1801 to 1809. As President, Jefferson bought the ... United States from France.

3.

Abraham Lincoln grew up in Kentucky in a log cabin. He couldn't go to school, so he taught himself. He became a lawyer. Friends called him "Honest Abe." As a delegate from Illinois, he served in Congress from 1847 to 1849. Lincoln was against slavery and gave some famous speeches about his ideas when he was running for the Senate.

In 1861 Abraham Lincoln became the sixteenth President of the United States. He wanted the states of the Union to work together as one country, but he had to lead the North against the South in the Civil War. Some people thought that Lincoln was too strong as President because he used power that the Constitution did not give him.

President Lincoln freed the slaves with the Emancipation Proclamation. He had a plan to bring the South back into the Union after the Civil War, but he couldn't carry out that plan because he was assassinated. In 1865 an actor named John Wilkes Booth shot Abraham Lincoln.

4.

John F. Kennedy was President for only three years, from 1961 to 1963, but his personality and ideas changed America. He was both the first Roman Catholic and the youngest President in the history of the country. He set clear goals for America. For example, he promised that the United States would land a man on the moon before 1970.

Kennedy supported the ideas of Martin Luther King, Jr. and fought for civil rights, fair housing, and programs to stop poverty. He asked Congress for more money for education and medical care for elderly people.

Kennedy was against Communism. For example, when the Soviet Union put missiles in Cuba, he sent U.S. ships to surround the island. But he believed that the best way to fight Communism was not by sending armies but by attacking poverty and injustice. He organized the Alliance for Progress to help the countries of Latin America. He started the Peace Corps and sent Americans to over sixty countries in Africa, Asia, and South America. These young volunteers worked and lived with the people, built schools, and taught farmers more modern methods.

Kennedy was a man for the future. He worked to stop the testing of nuclear weapons. But on November 22, 1963, he was assassinated.

The four selections in this "Expert Game" are from *English Through Citizenship, Intermediate Level A*, published by Delta Systems, Inc.

INSTRUCTIONS

1. Write the titles of the three to six articles or stories on the chalkboard. As a prereading exercise, discuss with the class what each reading might be about.

2. Each group of Ss receives the same set of readings, a different selection for each group member. Allow time for Ss to read their selections and perhaps take notes. Give individual help as needed.

3. Ss work in their groups to teach one another their material. In turn, they summarize the main ideas, answer questions, and perhaps even lead a brief discussion on the topic. Listeners take notes.

4. After a specified length of time, collect the reading cards. Reconvene the class to give an oral "quiz." Ask one or more questions about the information in each selection. Individually, Ss write answers in a numbered list, referring to their notes (but not the original articles or stories). For example, at a high beginning or low intermediate level, you might make statements about the information. After each number on their paper, Ss write the name of the president the statement is about. Here are sample items:

    1. He was the first president of the United States.

    2. He was president during the Civil War.

    3. He was the first Roman Catholic and the youngest president of the United States.

    4. He was a scientist, an inventor, a philosopher, and an architect.

    5. His ideas, such that "all men are created equal," became basic principles of U.S. government.

    6. He freed the slaves with the Emancipation Proclamation.

    7. He organized the Alliance for Progress and the Peace Corps.

    8. He was a military leader in the Revolutionary War.

    Alternatively, you might want to have the members of each group take the quiz cooperatively, making one answer list for the team.

5. Correct the quiz with the class by repeating the questions and having Ss tell their answers. Discuss difficult points, asking the S "experts" on each topic for verification if necessary. Collect scores.

"Famous People in History" is a good topic for an expert game because it is easy to find short biographical readings—even in simplified English.

LEVELS = HIGH BEGINNING TO LOW ADVANCED
↙ SUGGESTIONS FOR ADAPTATION ↘

⇓ In lower-level classes, divide the class into groups of only three or four Ss-- so that there will be fewer different articles or stories to deal with. In higher-level classes, allow Ss to be "experts" on up to eight different readings. Have fewer groups with more Ss in each.

⇔ Adapt the quiz questions to the writing abilities of the group: use *true /false* questions for lower levels and essay-type questions for higher ones. *Examples:*

- <u>T / F</u>: *Abraham Lincoln freed the slaves with his Emancipation Proclamation.*
- <u>Essay</u>: *What was the effect of Lincoln's Emancipation Proclamation?*

⇔ Instead of giving an oral quiz, hand out a list of questions (in one or more forms) or write the questions on the board; Ss work alone or in small groups to answer them.

POSSIBLE VARIATIONS

⇔ Instead of presenting their summaries in small groups, Ss circulate around the classroom teaching their area of expertise to any classmates who need that information. For example, if their are six different articles or stories, Ss seek out five classmates responsible for five different readings. To prepare for the quiz to follow, they take notes on these areas.

Sally Ride

Susan B. Anthony

Margaret Thatcher

Mother Teresa

OTHER AREAS OF APPLICATION: Any informational content area of interest to Ss and/or based on the theme of the textbook chapter being covered at the time, such as health, holiday customs, what to do in emergencies, resources for students at your particular school, health advice, etc.

"T he Expert Game" involves a variety of high-level comprehension and communication skills: reading for meaning, summarizing, speaking ("presenting" information) effectively, taking responsibility for understanding the presenter, learning content, and answering test questions. Here is another way to structure the activity.

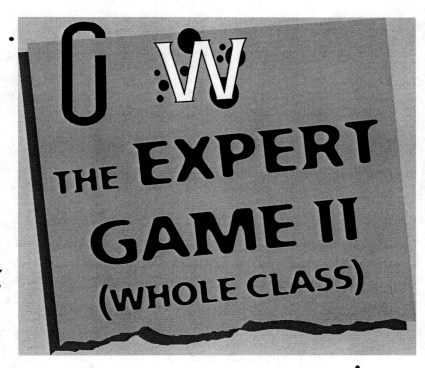

⇒ **SPECIFIC CONTENT OF THESE INSTRUCTIONS: Work improvement.**

⇒ **MATERIALS: Short articles at students' reading level--two copies of each, half as many articles as there are students--pasted up on index cards of two different colors; writing paper.**

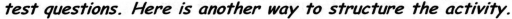

Planning

No football coach would think of sending his team into a contest without a game plan. Such a plan is not sacred; indeed, it is almost sure to be modified as the game progresses, but it is important that there be such a plan at the outset. You need a game plan for your day, and for your week. Otherwise you'll allocate your time according to whatever happens to land on your desk. Other people's actions will determine your priorities. And you will find yourself making the fatal mistake of dealing primarily with problems rather than opportunities. In planning your time, a general schedule of your day, with particular emphasis on the ...

Copies

Excessive record-keeping is a symptom of insecurity.

Figure out how often you use the various kinds of material you file. Take each category, and ask yourself, "What is the worst thing that could happen if this file didn't exist?" You'll find that most of the time the answer is "Nothing." If you really needed the information, it probably could be located elsewhere in the company in someone else's file. Or a phone call would do the trick, or you would get along fine without it.

This is not to say that comprehensive files ...

Crisis

Whenever you are faced with a crisis, ask yourself, "What can I do to prevent this crisis from recurring?"

Many of the crises that arise in business or in personal life result from failure to act until a matter becomes urgent, with a result that more time is required to do the job. For example, if you are behind schedule with an annual printing job, you can't mail your manuscript to the printer, you must hand-deliver it; you can't wait for him to deliver proofs, you must have someone pick them up; you can't delegate various tasks in connection with the job, you must do them yourself in order to gain time.

Whenever such a crisis occurs write yourself a note and put it in your "Future" file to appear on the date you should start working.

No

Of all the time-saving techniques ever developed, perhaps the most effective is the frequent use of the word no.

You cannot protect your priorities unless you learn to decline, tactfully but firmly, every request that does not contribute to the achievement of your goals.

The tendency of many time-pressured people is to accept grudgingly new assignments in vol...

Planning

No football coach would think of sending his team into a contest without a game plan. Such a plan is not sacred; indeed, it is almost sure to be modified as the game progresses, but it is important that there be such a plan at the outset. You need a game plan for your day, and for your week. Otherwise you'll allocate your time according to whatever happens to land on your desk. Other people's actions will determine your priorities. And you will find yourself making the fatal mistake of dealing primarily with problems rather than opportunities. In planning your time, a general schedule of your day, with particular emphasis on the two or three major things you would like to accomplish. One of those major things should be planning on a major project or on some task that will carry you closer to one of your lifetime goals. On Thursday or Friday do the same thing for the next week.

Remember, there is no more productive use of time than planning ahead. Studies prove what common sense tells us: the more time we spend in advance planning on a project, the less total time is required for it. Don't let today's busy work crowd planning time out of your schedule.

Copies

Excessive record-keeping is a symptom of insecurity.

Figure out how often you use the various kinds of material you file. Take each category, and ask yourself, "What is the worst thing that could happen if this file didn't exist?" You'll find that most of the time the answer is "Nothing." If you really needed the information, it probably could be located elsewhere in the company in someone else's file. Or a phone call would do the trick, or you would get along fine without it.

This is not to say that comprehensive files aren't useful, but the question is whether they are useful enough to justify the amount of time and effort that goes into keeping them current. Estimate the amount of time spent filing such things as old company house organs, routine memoranda, information copies of other people's memos, and so on, and ask yourself if the company wouldn't be ahead if the same number of hours were put into something directed toward achieving your primary goal.

Crisis

Whenever you are faced with a crisis, ask yourself, "What can I do to prevent this crisis from recurring?"

Many of the crises that arise in business or in personal life result from failure to act until a matter becomes urgent, with a result that more time is required to do the job. For example, if you are behind schedule with an annual printing job, you can't mail your manuscript to the printer, you must hand-deliver it; you can't wait for him to deliver proofs, you must have someone pick them up; you can't delegate various tasks in connection with the job, you must do them yourself in order to gain time.

Whenever such a crisis occurs write yourself a note and put it in your "Future" file to appear on the date you should start working.

Failure to start early enough is only one cause of crises. Others include misunderstandings due to unclear communications, lack of periodic status reports that can serve as an early-warning system, failure to follow through after delegating, and failure to make contingency plans.

Analyze each crisis, and see if you can devise ways of preventing a repetition. You'll find you can save enough time and energy to enable you to deal effectively with those relatively few cases in which circumstances totally beyond your control make it necessary to push the panic button.

No

Of all the time-saving techniques ever developed, perhaps the most effective is the frequent use of the word no.

You cannot protect your priorities unless you learn to decline, tactfully but firmly, every request that does not contribute to the achievement of your goals.

The tendency of many time-pressured people is to accept grudgingly new assignments in volunteer organizations, new social obligations, new chores at the office, without realistically weighing the cost in time. Such people worry about offending others--and "wind up" living their lives according to other people's priorities.

At work, of course, you cannot always turn down the request that you take on a job that you think a waste of time. But you can win a good percentage of the time if you try. Point out to your boss that the new task will conflict with higher-priority ones and suggest alternatives. If your boss realizes that your motivation is not to get out of work but to protect your time to do a better job on the really important things, you'll have a good chance of avoiding unproductive tasks. But you have to speak up.

These reading selections are pages from the book *Getting Things Done*, by Edwin C. Bliss.

Doing Without the Photocopier

INSTRUCTIONS

1. Keeping pairs of cards in order, count out the number of index cards corresponding to the number of Ss present. Shuffle these and give one card to each S.

2. Ss circulate around the classroom to find the S with the identical article. Meanwhile, list consecutive numbers on the chalkboard with space to write after each number. The number of items listed should correspond to the number of pairs of Ss (1-?).

3. Ss work in pairs to take notes on, outline, or summarize their article in writing. Circulate to give necessary help. (You can collect and comment on and/or grade these after the game.)

4. As soon as they have finished, S pairs write their names and the name of their article after one of the numbers in the list on the board. Pairs that have worked quickly can use the extra time to practice retelling the main ideas of their article to each other, you, or other Ss who have finished.

5. Using only their notes, outlines, or summaries, S pairs in turn (as listed in order on the board) explain the information from their aricle to the class. There are several ways they might work together:

   - One S can tell the main ideas, and the second can reiterate them, emphasizing the most important points.

   - One S can tell some of the main ideas, and the other can complete the summary.

   - One S can summarize the article, and the second can answer questions.

6. Including the names of the speakers and the article, each listener takes his/ her own notes in a numbered list. *Example:*

Notes on the Expert Game:
Saving Time at Work

1. *Victor + Denise: Copies: Don't keep too many records, organize copying and filing according to time priorities, get rid of papers that could be located elsewhere if needed.*

2. *Chien + Noriyuki: Crises: Occur because people wait until last minute to do necessary things. Plan ahead. Learn from each crisis. Keep it from happening the next time.*

3. *Adar and Mohammed: Planning: Plan every day and every week. It's the most productive use of time and saves working time. Set priorities and stick to them.*

4. *Akiko and Tina: How to Say "No": Learn to say no according to your priorities—otherwise you'll get off track and work and live for others.*

Short articles on how to do practical things make good materials for "The Expert Game."

LEVELS = HIGH BEGINNING TO LOW ADVANCED
↙ SUGGESTIONS FOR ADAPTATION ↘

⇓ For lower-level groups, provide articles in simplified English. Advanced Ss can handle authentic readings that are reasonably short.

⇑ In small high-level groups, give each S a different article for which s/he is alone responsible.

⇓ Help lower-level Ss take notes on the main ideas by asking questions about each speech. After each of the first few speeches, write notes on the board from S suggestions.

⇔ Adapt the quiz questions to the writing abilities of the group: use *true/ false* questions for lower levels and essay-type questions for higher ones. *Examples*

- *T / F: Office workers should make and file as many copies of everything as possible.*
- Essay: How can we reduce the amount of paperwork in an office?

⇔ Instead of giving an oral quiz, hand out a list of questions (in one or more forms) or write the questions on the board; Ss work alone or in small groups to answer.

POSSIBLE VARIATIONS

⇔ Instead of giving the same complete article to two Ss, divide each article into two parts: paste up the beginning on a card of one color and the end on a card of another color. (Alternatively, divide the article into three or four parts and paste up on three or four different-colored cards.) Ss skim their section for the topic before they look for their partners or groups.

⇔ Instead of handing out parts of articles, distribute questions on cards of one color and the corresponding answers on cards of another (e.g., interview questions and answers from a news magazine; problems and solutions from a newspaper advice column). After they have found their partners, Ss work together to summarize the question and the answer. One S in each pair presents the question or problem to the class. The class predicts the answer or solution or offers an original one. The second S summarizes the "real" (the original) response.

*O*THER AREAS OF APPLICATION: Any informational content area of interest to Ss and/or based on the theme of the textbook chapter being covered at the time, such as how-to advice, facts about consumerism, holiday customs, places to visit in your city, and more.

QUIZ BOARD GAMES

In many courses, including language courses, teaching and learning content (information about one or more subjects) is an important goal. Instead of relying solely on traditional lecture and reading methods, you can increase learners' motivation by presenting facts in the form of competitive and cooperative games.

⇒ **SPECIFIC CONTENT OF THESE INSTRUCTIONS:** General science (Tic-Tac-Toe game).

⇒ **MATERIALS:** For each pair or group of players, a set of 15-30 question cards with the answers on the back--each set of a different color; writing paper or a tic-tac-toe grid for each pair or group.

1. What instrument uses light rays to magnify small things up to 2500 times?
   a. a microscope
   b. a stalactite
   c. an X-ray machine

2. What U.S. government agency is in charge of space exploration?
   a. NASA
   b. the IRS and INS
   c. HUD

3. What is a star pattern like the "Big Dipper" called?
   a. An astronaut
   b. A meteoroid
   c. A constellation

4. What happens during an eclipse of the sun?
   a. The moon moves in front of the sun.
   b. It rains.
   c. The sun moves to the other side of the earth.

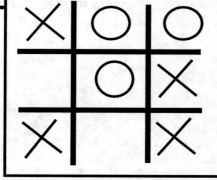

Some sample materials—a board and question cards—for a *Quiz Board Game*.

INSTRUCTIONS

1. Prepare sets of question cards beforehand. The questions in each set can be on a separate topic, such as *general science, biology, zoology, math, astronomy, computer technology*, and the like, or each set can be a subset of questions related to one general topic. For instance, (e.g., *astronomy = the universe, the solar system, the stars, space exploration*, etc.).

2. To play a sample game with the whole class, draw a large *Tic-Tac-Toe* grid on the board. If necessary, review the rules of *Tic-Tac-Toe*--and perhaps game strategy. Explain that in the quiz version of this game, players must answer a question correctly before they can put an X or an O on the board.

3. Divide the class into two teams, one represented by X's and the other represented by O's. Ask each team in turn--either a specific S or the team as a whole--a question from one set of cards. If the answer is correct, the player who answered places that team's symbol in one of the squares in the *Tic-Tac-Toe* grid. To win the game, a team must place three symbols on the grid in a horizontal, vertical, or diagonal row. You might provide a prize that can be shared to the winning team.

4. Divide the class into groups of three (two players and a questioner, five (two teams of two Ss each, plus a questioner), or seven (two teams of three Ss each, plus a questioner). Each group receives a set of question cards and produces a tic-tac-toe grid on a piece of paper.

Teaching Tips

If you wish to reuse these grids, you can prepare them on card stock and provide markers of two types--such as red and black checkers, two colors of beans, etc.--to each group. Ss place markers on squares instead of writing X's and O's.

5. Ss play the *Tic-Tac-Toe* game as described above. The questioner asks each player or team in turn a question (checking the answer on the back of the card and providing the correct answer, if necessary), enables them to places their X's or O's (or markers) on the grid, and gives a small prize to the winner or the winning team. For the next game, the questioner trades roles with one of the players. Play continues for several rounds. After all the questions have been asked at least once, the group trades cards with another group.

Reconvene the class and collect the sets of question cards. Review the information by playing the game with the whole class several more times, as described in Step 3 above, and/or by giving an oral or written "quiz" on the material.

All kinds of board game formats can be used in connection with question-and-answer games.

LEVELS = HIGH BEGINNING TO LOW ADVANCED
↙ *SUGGESTIONS FOR ADAPTATION* ↘

⇔ Adapt the content of questions to both Ss' level of language ability and knowledge. For lower levels, provide *yes / no* or *true / false* questions. *Examples:*

- *Yes / No: Do scientists use a telescope to magnify small things?*
- *True / False: The sun is made of hot gases.*

For high beginning and intermediate groups, provide multiple choice questions, such as the questions on the sample cards on page 70. For advanced groups or for review, you can provide *wh*-questions, such as:

- *What instrument uses light rays to magnify things up to 2500 times?*
- *What U.S. government agency is in charge of space exploration?*
- *What happens during an eclipse of the sun?*

POSSIBLE VARIATIONS

⇔ Instead of a tic-tac-toe grid, supply (or have Ss draw) a grid for another board game with reasonably simple rules, such as *Checkers, Hex, Nine Men's Morris*, and the like. Players must answer a question before they are allowed to make a move.

⇔ Instead of a board game, devise a grid patterned after a TV quiz show, such as "Jeopardy." Simplify and adapt the TV game rules for the classroom--either whole class or small groups.

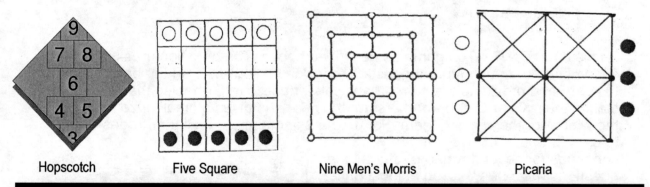

| Hopscotch | Five Square | Nine Men's Morris | Picaria |

Every culture has its own game board patterns, with various kinds of rules for play.

OTHER AREAS OF APPLICATION: Any content area in which facts are taught, such as history, government, geography, economics, literature, art, music, math, and many more.

There are many ways of providing insight into North American culture that involve language. In addition, discussions of American culture often lead to cross-cultural comparisons (of similarities and contrasts) that help learners to understand their own and others' perceptions of "reality." To stimulate critical and creative thinking, try this simple "Game of Wisdom" in one or more of its many varieties.

⇒ **SPECIFIC CONTENT OF THESE INSTRUCTIONS: Proverbs.**

⇒ **MATERIALS: Writing paper; a class set of index cards (at least one for every student), with half of a common proverb printed on each card. It is best to use one color card for all the beginnings and another for all the endings.**

A sample "Wisdom Game" based on American proverbs.

INSTRUCTIONS

1. Each S receives a card with half a proverb on it. Ss walk around the classroom looking for their partner—the person with the matching card. Pairs discuss the meaning of their proverb (situations to which it might apply).

2. Beforehand or meanwhile, list paraphrases, one for each proverb, on the chalkboard. When the class reconvenes, read these aloud one at a time. Each pair of Ss in turn shows and explains the corresponding proverb to the class. Write, or have Ss write, each proverb on the board after the paraphrase. During this time, Ss can copy the proverbs in their own individual lists.

3. Erase or cover the paraphrases. Describe situations. For each, Ss read aloud the appropriate proverb from the board. For example, *I can't complete this activity by myself; but with a partner, I'll be successful.* = Proverb: *Two heads are better than one."*) For review, erase the proverbs on the board and have the class recall them—without looking at their lists—from the paraphrases or descriptions of situations.

4. To use this activity for a writing lesson, with the class analyze the sentence types represented by the proverbs (e.g., simple, compound, complex) or the sentence patterns (e.g., Subject-Verb-Object-Place-Time). Ss write translations of proverbs from their own culture, perhaps identifying the sentence types or patterns. They put these on the board for the class to read and discuss. The class groups those proverbs that have essentially the same—and/or opposite—meanings.

Teaching Tips

You may want to assign specific proverbs from the previously-discussed list of American proverbs to groups of Ss, having them concentrate on proverbs from their cultures with similar (or contrasting) meanings.

You never really know your friends . . .

Ultimately, takers lose . . .

Do what you know is right, . . .

If things are not as bad as they seem, . . .

A friend in need . . .

Wherever you go, . . .

. . . is usually a pest.

. . . the world's worst drivers will follow.

. . . and givers win.

. . . but try not to get caught at it.

. . . why do they seem so bad?

. . . until you take a vacation together.

Some more modern "sayings" for the *Game of Wisdom*. If possible, encourage students to come up with their own bits of "wisdom." Print their thoughts on cards for your next game.

LEVELS = HIGH BEGINNING TO LOW ADVANCED
↙ SUGGESTIONS FOR ADAPTATION ↘

⇓ For lower levels, reduce the number of proverbs and matching paraphrases. You might provide two copies of the parts of each proverb, so that there are only 1/4 as many proverbs as Ss.

⇑ After completing the paraphrase list, higher-level Ss can write about one or more of the proverbs, describing a situation in which it might apply, telling their opinion (*Is the proverb true for you? Why or why not?*), and so on.

POSSIBLE VARIATIONS

⇔ Instead of writing paraphrases on the board as in Step 2 above, provide a previously-prepared list of paraphrases, one copy to each pair of Ss. S pairs circulate around the class together, looking for the proverbs that match the paraphrases on their list. They write each proverb after the corresponding paraphrase or on paper (so that the lists can be reused). The first few pairs of Ss to complete the task receive small prizes. While others are still working, they list the proverbs on the board in preparation for follow-up activities.

⇔ Instead of matching cards with parts of proverbs on them, Ss match proverb cards with paraphrase cards. When they have found their partners, they write their proverb on the board. The class reconvenes. Each S pair in turn reads aloud the paraphrase. The class identifies the matching proverb.

OTHER AREAS OF APPLICATION: Fortunes (from fortune cookies or made up), sayings from the Peter Principle (e.g., *If there is a warranty on it, a product will not break. It will break as soon as the warranty expires*), famous or witty quotations (See the books of *Potshots* by Ashley Brilliant.), truisms or philosophical statements. Books in the gift section of bookstores and magazines like *Reader's Digest* are a good source of these.

Teaching Tips

If the sentences are in simple language, there is no need for paraphrases. Simply provide sentence halves to Ss, have them find their partner and list their sentence on the board, and follow the relevant instructins from Steps 4 and 5.)

GREETING CARD CULTURE

Most language has cultural overtones, and language courses are a fitting environment for lessons on and discussions of culture. How can you make use of all the objects and cultural communications you receive and collect through the years? Use them for language lessons!

⇒ **SPECIFIC CONTENT OF THESE INSTRUCTIONS: Greeting cards for American special occasions and holidays.**

⇒ **MATERIALS: A set of real greeting cards received on various occasions--birthdays, graduation, weddings, births, anniversaries, holidays, etc.**

Teaching Tips

When you are collecting cards for this activity, check that the messages written by hand on the cards are not too personal for you to share with the class. If they are, don't include those cards.

Some examples of typical greeting cards—outsides and insides—collected over the years.

INSTRUCTIONS

1. Ss work in pairs or small groups to prepare a chart to fill in. The first (narrow) column is for card numbers. Here are suggested questions for headings for the columns, with some sample answers:

Card Number	What occasion is the card for? (What kind of card is it?)	What kind of person sent it? To whom?	Describe the pictures.	What typical special occasion words appear on it?	Would you give or send this card? If so, to whom?
1	Sickness—It's a Get-Well card	A friend—to a sick person	Cartoon animal. A chicken in bed	Get-Well, advice, feel better	Probably—if they can get the joke.
2	A birth announcement	New parents—to relatives & friends	Cute, little pictures of a child, animals . . .	6 pounds, 11 ounces, 19 1/2 inches, proud parents	Not yet—maybe in ten years or so.
3	The anniversary of a marriage	A friend—to a married couple	Cute drawing of two cats	anniversary, together, husband, wife, wonderful	Yes—to my aunts and uncles or parents.
4	A birthday	Relative—to his sister.	A photo of a road	50-year-old, age, happy birthday	No. It's mean to joke about age.

Teaching Tips

It may help Ss to follow instructions correctly if you provide paper of the appropriate size to the task. In this case, 8 1/2 x 14 (legal size) paper would be useful. Ss should turn it lengthwise to make their charts.

2. Show a few cards to the class and discuss them. Here are some more possible answers to the above questions:

   - *a birthday, a wedding, a bar mitzvah, an anniversary, a divorce, an invitation to a party, Valentine's Day, Christmas, New Year's, a religious occasion, . . .*

   - *humorous, serious, sentimental, religious, thoughtful, friendly, warm, mean, . . .*

   - *a friend, a relative, a doctor, a colleague, a company, . . .*

   - *a rose in a garden, a bride and groom on a cake, a garden, children, animals*

   - *last a lifetime, happiness, loving pride, special thought, in sympathy, the best, ...*

   - *no, because I don't have a daughter; yes, because I like to send funny cards to my friends*

3. Pass the cards around the class. Ss work together to fill out their charts; these can be collected for comment after the activity.

4. After a specified period of time, collect the cards. Show each card; groups in turn tell their answers. Encourage questions, comments, and discussion about the occasions and related customs.

5. For follow-up, redistribute the cards; have each S write a "personal journal" type entry of associations and thoughts that their card elicits. They can share these with the class and/or hand them in for comment.

LEVELS = HIGH BEGINNING TO INTERMEDIATE
↙ SUGGESTIONS FOR ADAPTATION ↘

⇓ For lower levels, provide only a small number of cards--only with simple messages. Limit the number of occasions. Write these on the board and provide a brief introduction to each before beginning the activity.

⇑ Provide higher levels with a large variety of cards, including some that illustrate the "culture" of American humor. Before beginning the activity, Ss sort the cards into categories--such as cards for children, adults, relatives, friends; religious cards, serious or sentimental cards, humorous cards; or some other classifications of their choice. As described above, they work with the cards in small groups; when they have finished, each group exchanges its set of cards with another group.

POSSIBLE VARIATIONS

⇔ Ss provide greeting cards or post cards from their own cultures. Other Ss try to guess what kind of cards they are, what they show, what kind of person might have sent it, etc.

OTHER AREAS OF APPLICATION: For greeting cards: fall and winter holidays, spring and summer holidays; picture postcards (Examples of possible questions: Where is the card from? What does the picture show? Who wrote the message? To whom? What is the main idea of the message?); business cards (Examples of possible questions: Whose card is it? What is his/her position? With what kind of company? To whom might s/he give out this card?)

Some examples of other kinds of cards useful in teaching and learning about culture.

What Comes Next?

ARE YOU READY TO GO BEYOND THE IDEAS IN <u>DOING WITHOUT THE PHOTOCOPIER</u>?

Some years ago, when I finished writing **DOING WITHOUT THE PHOTOCOPIER,** I assumed (mistakenly) that I would never have to think of any more creative generic ideas—that the 26 concepts A to Z would suffice for the rest of my (and my colleagues') teaching career—and beyond to teacher-training workshops in the 21st century.

This second volume, **STILL DOING WITHOUT THE PHOTOCOPIER: FROM AA TO ZZ,** addresses some recent changes and expected future changes in language education—the ever-shrinking budgets for classes, materials, and curriculum; the in-fights and outside competition for "territory" (tax money) and students; the accessibility (or expectation of the availability) of useful multi-media equipment—including computers and the Internet; the increasing demands of motivated students for efficient and effective language education (not wasteful of their time or money); experienced and perhaps burned-out faculty's need for change and novelty, and a myriad of other developments, most of which are not yet identified.

To meet the needs and interests of teachers and learners growing more ever more sophisticated and demanding, Ideas AA-ZZ are even more generic, complete, and flexible than those in the previous book. A condensed Table of Contents appears on the following pages.

Again, I hope I've helped. Again, I invite your reactions, comments, and suggestions.

E.K.-R. Culver City, California

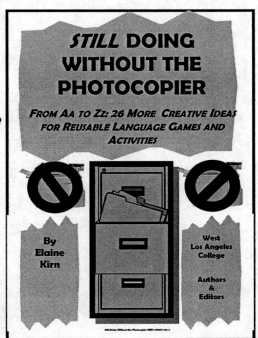

STILL DOING WITHOUT THE PHOTOCOPIER

FROM AA TO ZZ: 26 MORE CREATIVE IDEAS FOR REUSABLE LANGUAGE GAMES AND ACTIVITIES

By Elaine Kirn

West Los Angeles College

Authors & Editors

**Still Doing Without the Photocopier
ISBN 1-891077-24-4**

STILL DOING WITHOUT THE PHOTOCOPIER
CONDENSED TABLE OF CONTENTS

SKILLS FOCUS: GRAMMAR & PHRASING

IDEA JJ: VIDEO GRAMMAR-WRITING"

Instructions: How to Apply Grammar Patterns and Rules to Simulated "Real Life"

IDEA KK: CLASSIFYING CLUTTER

Instructions: How to Collect and Make Use of Cluttered Visuals in Grammar Activities

IDEA LL: ODDS & ENDS

Instructions: How to Make Productive Educational Use of Collected Odds & Ends

IDEA MM: QUICK—WHAT'S THE QUESTION?

Instructions: How to Create and Use Quiz Answer Boards for Question Formation

SKILLS FOCUS: ORAL LANGUAGE

IDEA NN: FOCUS ON SENTENCE FOCUS

Instructions: How to Create and Use Statement Strips with Focus

IDEA OO: OPEN-ENDED OPENERS

Instructions: How to Create and Use an Open-Ended Question Card Deck

IDEA PP: SPONTANEOUS SPEECH & SPEECHES

Instructions: How to Prepare and Use Materials for Speech and Speeches

IDEA QQ: BRAINSTORMING

Instructions: How to Prepare and Make Productive Use of Materials for Brainstorming

IDEA II: PUT IT IN CONTEXT

Instructions: Teaching & Learning Vocabulary from Context